HEAT

ALSO BY BILL STREEVER

Cold: Adventures in the World's Frozen Places

HEAT

Adventures in the World's Fiery Places

BILL STREEVER

Little, Brown and Company

New York Boston London

Little, Brown and Company
Hachette Book Group
237 Park Avenue, New York, NY 10017
littlebrown.com

First Edition: January 2013

Little, Brown and Company is a division of Hachette Book Group, Inc. The Little,
Brown name and logo are trademarks of Hachette Book Group, Inc.

The publisher is not responsible for websites (or their content) that are not owned
by the publisher.

The Hachette Speakers Bureau provides a wide range of authors for speaking
events. To find out more, go to hachettespeakersbureau.com or call (866) 376-6591.

Library of Congress Cataloging-in-Publication Data
Streever, Bill.
 Heat : adventures in the world's fiery places / Bill Streever.
     p. cm.
 Includes index.
    ISBN 978-0-316-10533-0
 1. Arid regions—Description and travel. 2. Heat. I. Title.
 GB611.S77 2013
 551.41'5—dc23   2012020861

10 9 8 7 6 5 4 3 2 1

RRD-C

Printed in the United States of America

For my companion and wife, Lisanne Aerts,
who is always eager to go somewhere new
and unusual no matter what the temperature
may be. And for my son, Ishmael Streever, who
will eat the hottest of peppers without hesitation
in exchange for a nominal fee.

Heat can also be produced by the impact of imperfectly elastic bodies as well as by friction. This is the case, for instance, when we produce fire by striking flint against steel, or when an iron bar is worked for some time by powerful blows of the hammer.

—Hermann von Helmholtz,
On the Conservation of Force (1862)

Eating coals of fire has always been one of the sensational feats of the Fire Kings, as it is quite generally known that charcoal burns with an extremely intense heat.

—Harry Houdini,
The Miracle Mongers: An Exposé (1920)

CONTENTS

A NOTE ON DEGREES FAHRENHEIT

Throughout *Heat*, I use degrees Fahrenheit because it is the temperature scale most familiar to most readers. To convert from degrees Fahrenheit to degrees Celsius (or centigrade), subtract 32 and then multiply by 5/9. Here are a few examples, crossing the range of temperatures explored in *Heat*:

Degrees Fahrenheit	Degrees Celsius
0	−18
32	0
80	27
125	52
212	100
2,200	1,204
11,000	6,093
7 trillion	Just under 4 trillion

A CANDLE'S FLAME

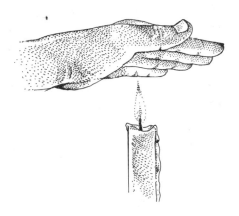

On a table in my suburban Anchorage home: a stubby red candle, matches, an electronic thermometer, a bowl of snow, and my ungloved left hand.

I look at the palm of my left hand: for now unburned, soft and uncalloused, scarred from a long-ago accident with a knife. On the back of my hand: fine blond hairs and early wrinkles of aging skin.

I strike a match, hold it while the flame stabilizes, and touch it to the candle's wick.

I move the back of my hand through the flame and smell melting hair. Hair is keratin, the same stuff in feathers and hooves and fingernails and baleen, a complicated protein, its long molecules coiled and folded upon themselves. The smell is that of protein molecules uncoiling and unfolding, coming unglued, losing their three-dimensional structure, denaturing.

Firefighters sometimes talk of skin melting. A Chicago firefighter struggled to carry a woman from a burning house. "She was slipping out of my hands because my skin was melting," he later told an interviewer. "On my left hand, the melted skin had fused together my fingers."

I pass my palm quickly through the flame. I feel a mild sensation of heat. One of the lessons of heat: duration of contact matters.

From Herman Melville's Mr. Stubb, lover of rare meat and second mate aboard the *Pequod* during her pursuit of Moby Dick, on how to cook a whale steak: "Hold the steak in one hand, and show a live coal to it with the other."

Two centuries before Melville, scientists tried to understand heat. The German scientists Johann Joachim Becher and Georg Ernst Stahl promoted the phlogiston theory. Flammable materials contained phlogiston, a colorless substance with neither taste nor odor. Burning removed phlogiston. A burning candle became dephlogisticated. A steak held to a live coal long enough could become dephlogisticated. My hand, held in the flame long enough, could become dephlogisticated.

A century later, the great French scientist Antoine Lavoisier, the man who named oxygen and hydrogen, dephlogisticated the phlogiston theory. He argued that heat was a subtle fluid. He called the fluid "caloric."

Antoine Lavoisier was as wrong as Becher and Stahl.

Slowly, I move my hand, palm down, toward the flame. At a distance of six inches above the flame, the heat becomes uncomfortable. It becomes a threat. I pull back.

During Melville's lifetime, Michael Faraday played with candles. Faraday is best known for his experiments and discoveries in magnetism and electricity, but he cherished the burning candle as a lecture topic. In 1860 he published his lectures as *The Chemical History of a Candle*.

"There is no more open door by which you can enter into the

study of natural philosophy," Faraday wrote, "than by considering the physical phenomena of a candle."

With my electronic thermometer, I measure my candle's flame. It is an infrared thermometer, a pyrometer, capable of measuring heat from a distance, capable of seeing heat as the eye sees light. It looks beyond the visible spectrum, seeing wavelengths longer than the reddest of reds. The thermometer has the shape of a toy pistol, a gray phaser with a yellow trigger. Heat waves enter through the barrel, pass through a lens, and are focused on a thermopile. The thermopile converts heat waves to electricity. Circuitry converts the electricity into a digital readout. The digital readout says 1,350 degrees.

Faraday, in his candle lectures, described the candle flame as a hollow shell. Inside the shell there is no oxygen and therefore no fire.

I attach a steel-tipped probe to my thermometer, bypassing its optics. I push the probe into the unburning core for a reading of 1,200 degrees. At the tip of the flame, where oxygen meets hot gas, where ignition creates a yellow flare, I read 1,360 degrees, only 10 degrees off from the infrared reading. This is where I plan to bathe my hand, a five-second intentional exposure.

Twenty-three years after the publication of *Moby Dick* and seven years after Faraday died, Harry Houdini was born. Houdini grew up to become a Vaudeville star, an escape artist and illusionist, a film director, a skeptic, an exposer of lesser magicians, and a theatrical pyromaniac. Among his many offerings: instructions for cooking a steak. Houdini's instructions involved something more than a hot coal. To cook a steak, he required a cage. The cage should be made of iron, with dimensions of four feet by six feet. Its bars should be wrapped in oil-soaked cotton.

"Now take a raw beefsteak in your hand and enter the cage," Houdini wrote.

A stagehand ignites the oil-soaked cotton wrapped around the bars. Smoke and flames hide the performer from the audience. Inside, the performer hangs the steak from a hook above his head and then lies facedown on the bottom of the cage. The performer's cloak, including a hood, is made from asbestos fabric. The performer breathes through a hole in the bottom of the cage. The heat rises and cooks the steak.

"Remain in the cage until the fire has burned out," Houdini instructed, "then issue from the cage with the steak burned to a crisp."

I stare at the candle's brilliant yellow, its flickering blue, the molten wax at its base. From a society that promotes meditation: "Relax your body and mind as much as you can, and watch the candle flame. Do not strain your eyes while focusing on the candle flame, as the focusing you will be doing is in your mind, not in your eyes. Clear all of your thoughts, and focus them only on the flame."

Looking into the flame, I cannot clear my thoughts. I think of Faraday. He mentioned the "sperm candle, which comes from the purified oil of the spermaceti whale." During lectures, he ignited alcohol, phosphorous, zinc, hydrogen, a reed soaked in camphene, and the spores from club moss. He burned everything in reach.

John Tyndall was a friend and admirer of Faraday, a user of a primitive thermopile, and Faraday's successor as director of the Royal Institution. Tyndall was intimately familiar with Faraday's candle lectures. He discovered that carbon dioxide can trap heat in the atmosphere. And he knew that heat was not a substance. It was not a fluid, subtle or otherwise. It was an expression of molecular motion.

From Tyndall, in 1863: "The dynamical theory, or, as it is sometimes called, the mechanical theory of heat, discards the idea of materiality as applied to heat. The supporters of this theory do not believe heat to be matter, but an accident or condition of matter; namely, a motion of its ultimate particles."

Tyndall was a serious man, a man of science, and we have no record of him bathing his hand in a candle's flame. The title of his 1863 book: *Heat: A Mode of Motion*.

Quickly this time, I move my palm down into the candle's flame. I feel intense pain. I hold a world of pain in the palm of my hand.

Doctors sometimes use numerical pain scales, instructing patients to self-assess. Zero is no pain, five is moderate pain, and ten is the worst imaginable pain.

In the palm of my hand, I hold an eleven.

After two seconds, I am standing in a mild panic. By three seconds, the only way I can continue is to grasp my left wrist with my right hand. I command my right hand, well clear of the candle's flame, to hold my left hand in place.

At five seconds, I pull away from the candle. I plunge my hand into the bowl of snow. The pain drops from somewhere near eleven to something manageable. A nine. A seven. A three. I sit down. I pant. I stare at the candle flame. And I realize that today is the day I begin to understand heat.

HEAT

Chapter 1
RAVING THIRST

Heat, from a human perspective, is more than just a matter of high temperatures in the shade. Two doctors working in the 1950s watched marines in the heat and humidity of Paris Island, North Carolina. They wrote of "disabling heat" at temperatures in the eighties. Young marines wore fatigues, helmets, cartridge belts, bayonets in scabbards, and boots. They carried full canteens, rifles, entrenching tools, and twenty-two-pound packs. Marching in the sun, running in the sun, low-crawling in the sun, digging foxholes in the sun, they complained of cramps and exhaustion. Young marines occasionally collapsed.

The doctors saw that temperature alone could not predict heatstroke and heat exhaustion as well as what became known as the Wet Bulb Globe Temperature. When a marine recruit suffered in the heat, he also suffered in the humidity and the sun. The Wet Bulb Globe Temperature combined the air temperature in the shade, the air temperature in the sun, and something called the "natural wet-bulb temperature." The air temperature in the sun is measured with a black globe thermometer—a thermometer inside a blackened sphere of copper, left in the sun. The natural wet-bulb temperature comes from a thermometer with its bulb covered by a moistened sock. As the moisture evaporates, it removes heat. Increase the humidity and decrease the rate of evaporation, and in turn increase the wet-bulb temperature. Add a breeze to increase the evaporation rate, and in turn decrease the wet-bulb temperature.

Using the Wet Bulb Globe Temperature, the doctors figured out what every child knows: humid days feel hotter than dry days, and shaded ground is cooler than sun-bathed ground.

A flag system formalized their findings. A yellow flag warned against exercise in the sun: "Instruction involving strenuous physical exertion, such as an infantry drill, crawling on the ground, or bayonet training, will be given only in the shade." A red flag meant it was too hot to work: "During the time a red flag is displayed, recruits will be given only that type of instruction which can be given to men standing or sitting in the shade." Later a black flag was added: "All nonessential physical activity will be halted for all units."

My companion and I fly south to Las Vegas and rent a car. She drives while I stare at the sun-burdened landscape. Warning signs mark the entrance to Death Valley itself: "DANGER: EXTREME HEAT" and "LIMIT OUTDOOR ACTIVITY TO SHORT WALKS." If there were flags here, they would be black.

We park near the lowest point in Death Valley and walk into a canyon. We carry a gallon jug filled with water. We walk between vertical walls of white and yellow mudstone that leave the floor of the canyon in almost constant shade, and a breeze funnels between the walls. Still, the heat feels dangerous. It threatens us as though it has intentions, as though it wants to stifle breathing. The plan is a cautious walk, half an hour at the most, a stroll to get a feel for the land.

But that is not possible. The land lures us. This is a place that we must experience. Where we had planned to turn around, at the end of the canyon, we find a trail leading through badlands. We are in open sun now, in a place where the black globe thermometer temperature might approach the boiling point, but we are saved by a combination of low humidity and breeze and stupidity. We move up and down across low but steep hills that may once have been sand dunes. To our left, a red cliff towers. To our right, scorched hills stretch into the distance.

As a measure of the brutality of this land, cactus does not grow here. This place is too dry and too hot for cactus. But we come across occasional bushes, mostly desert holly, which is not a true holly at all. It is a chenopod, with leaves shaped like those of a holly, but paler.

The trail branches and then branches again. We find a sign with arrows that should point right and left, but the sign is lying on the ground, pointing nowhere. We find another sign, this one warning of abandoned borax mines, and we follow it, wondering what a borax mine might look like.

After an hour, our gallon of water has become a half gallon. The breeze has died.

I begin to wish that we had told someone where we were going.

Here is the irony of Death Valley: Its desiccated beauty was carved by water. As recently as 1976, a sudden flood destroyed a road that ran near here. Everywhere signs of erosion by flowing water prevail. This is not gentle erosion over time. This is sudden erosion, driven by water that appears from rare cloudbursts to rush through channels, carrying away dirt and stones and boulders and leaving behind dry channels called arroyos.

We walk down an arroyo now, comforted by occasional footprints as we move through our second hour.

With my infrared thermometer, I shoot a temperature in a tiny patch of shade: 112 degrees. I shoot another in the sun: 124.

In this heat, without water, with little shade, it is possible to die in as little as seven hours. Death would come from a combination of dehydration, heat exhaustion, and heatstroke. The blood's plasma volume would drop. The blood and the water bathing tissues and residing in bones would grow increasingly salty. Firing of nerve cells would become irregular as membranes changed and ion pumps—cellular pumps that maintain the interior chemistry of cells—failed.

When water for sweating is exhausted, when sweating stops, core temperature rises. The victim succumbs to heatstroke. As the core temperature reaches 107 degrees, the lining of the intestine fails, releasing fatty sugars into the bloodstream and triggering what amounts to septic shock. Among other things, the brain grows ill. In an end such as this, death comes from systemic failure, from a syndrome expressing in a hundred different ways the fact that the body no longer functions. A mere eight- or nine-degree increase in core temperature kills without mercy.

We move slowly, resting often, drinking in the landscape. The arroyo seems to loop back to the paved road where we parked our car, a path that will let us see the desert without backtracking. We find slivers of shade against pale cliffs of tortured mudstone that rise up from the arroyo. Ripples decorate the mudstone walls, more signs of flowing water. In places, mud liquefied by long-ago rains settled and was baked into flat shelves that make perfect benches in the shade. When we sit, the heat of the ground moves up through the seat of the pants.

We are down to a single quart of water. We worry that the arroyo may not lead to the road after all. We agree to turn around and retrace our path before drinking again, but then we press on.

A dry creek bed, known in this part of the country as a dry wash, intersects the arroyo's right bank, forming a confluence of dusty channels. Above it, I see a mine shaft. A mound of mine tailings stands just outside the opening of the shaft. The dry wash is steep, and the mine shaft is up an even steeper incline above the wash, a hundred yards away under full sun. But I want to see the shaft.

The shaft's opening is seven feet tall and three feet wide, the shape of an arching doorway, a tunnel entrance standing dark but surrounded by sunlit pale rock. The shaft itself runs straight back into the hill. I feel my way forward, blind in the shadows, shuffling my feet, and I am surprised when the tunnel ends. I stand in what feels like total darkness perhaps twenty-five feet from the entrance. Here, in the back of the tunnel, the heat is gone. The temperature is below ninety degrees. I stand in the darkness, savoring cool air.

Back outside, nearby, I find another tunnel. This one is cut through a hill, with openings on both sides, a would-be breezeway but for the total absence of air movement. In the middle, halfway between the openings, two rusted iron bed frames sit end to end. Miners must have slept here. Where they found water is not clear. Rainfall here averages two inches per year. Some years, no rain falls at all.

When they were not sleeping, the miners were busy. The hills around their breezeway are rotten with tunnels. In one, I find a half cup of guano piled on the floor. In another, I find distinct lines of salt, white crystals in the tunnel's wall. I lick the rock. It is not sodium chloride. It could be sodium borate. The miners sold the borate salts, a compound of boron used in fertilizers, insecticides, detergents, wood preservatives, and cement. Added to glass, boron prevents cracking from heat.

Added to some fabrics, it acts as a flame retardant. Added to iron smelters, it removes unwanted oxides. Added to nuclear reactor shields, it blocks neutrons. Added to humans in large-enough quantities, it causes headaches, vomiting, and eventually liver cancer.

The big borax operations in Death Valley were out near Furnace Creek, several miles from here, in the 1880s. Near Furnace Creek, miners dug for ore rather than the almost pure salts. The borax ore was boiled in vats with water and carbonated soda, and the solution was cooled in tanks wrapped in wet felt. Water evaporating from the felt sped the cooling process. When cooled to 120 degrees, borax salts crystallized.

Over a few decades, twenty million pounds of borax salts were hauled out in twenty-mule trains—two wagons full of borax and a water tank pulled by twenty mules or, more commonly, eighteen mules and two horses. The wagons had spoked wheels seven feet in diameter.

The distance to the nearest railroad: 165 miles.

During summers the cooling vats could not be cooled to 120 degrees. The borax ore miners took their summers off.

From a 1907 Death Valley newspaper advertisement: "Would you enjoy a trip to Hell? You might enjoy a trip to Death Valley! It has all the advantages of Hell without the inconveniences."

We make our way down the wash, back into the arroyo. I shoot the temperature of a rock in the sun: 129 degrees. We drink all but the last two tablespoons of our water, now as warm as tea, but abandon our plan to turn back. The road, we think, is just ahead, just around the next turn. My companion's face is flushed. A heat rash has blossomed on her legs. It has been three hours since we left our car.

The badlands that we move through are vegetated to the same extent that an asphalt parking lot is vegetated. We see, at times, the tracks of lizards, but never the lizards themselves. In places a black crust covers the white cliffs. Elsewhere gray-green veins run through the rocks.

The arroyo narrows into a gorge with walls too steep to climb. Now we walk in shade. We climb down rock ledges in the gorge, losing elevation. I feel ill, but the symptoms are not concrete. It is a general malaise, a feeling of exhaustion mixed with impatience.

At four hours, the arroyo returns us to pavement. We find our car. I shoot temperatures. The asphalt road registers 148 degrees. Inside the car, the driver's seat registers 180 degrees.

※

Antoine Lavoisier, French nobleman and brilliant chemist, promoter of the caloric theory of heat, the man who dephlogisticated the phlogiston theory of heat, realized that animals burned food in a reaction identical to that in which woodstoves burned firewood and in which lamps burned whale oil. "La respiration," Lavoisier wrote, "est donc une combustion." In English: "Breathing is thus a combustion."

Also from Lavoisier, with regard to animals that do not eat, that fail to find food: "The lamp would very soon run out of oil and the animal would perish, just as the lamp goes out when it lacks fuel."

Lavoisier knew, too, about the need for water. He convinced himself that his caloric theory explained sweating. In his mind, warmth meant that caloric was present, a subtle fluid, the embodiment of heat. Water dissolved caloric and carried it away through the skin as sweat, cooling the body. "It is not only by the pores of the skin that this aqueous emanation takes place," he wrote. "A considerable quantity of humidity is also exhaled by the lungs at each expiration."

His caloric theory was wrong, but he was right in believing

that water is lost as sweat and in the breath, and both contribute to cooling.

✳

Pablo Valencia probably never heard of Antoine Lavoisier, but he knew what it was to sweat. He was once a sailor on the Pacific Ocean, probably on vessels powered by a combination of wind and steam, but by the time he reached forty he had become a grower of watermelons and a prospector, five feet and seven inches tall, weighing 155 pounds. In 1905, the same year that a railroad company auctioned off land to create what would become Las Vegas, forty-year-old Pablo Valencia went into the desert just south of Las Vegas, a desert cooler than Death Valley, wetter, and less deadly.

It was a Tuesday, August 15. Pablo was with Jesus Rios, another prospector. The two men met William McGee, a scientist who mixed geology with anthropology and ethnology, who was camped at a place called Tinajas Altas, or "High Tanks," a natural trap for the little rain that fell each year. The prospectors were on horses. They carried six gallons of water, flour made from mesquite seeds or maize, bread, cheese, sugar, coffee, tobacco, and a combination of pressed alfalfa and rolled oats for the horses. They left McGee's camp at dusk. Sometime before midnight, Pablo and Jesus had covered thirty-five miles. Pablo sent Jesus back for more water, with plans to rendezvous a day later.

That rendezvous never happened. Jesus went back to McGee's camp for water and returned a day later to report that Pablo was missing. McGee mounted a search of sorts, doing the best he could in such remote land.

Temperatures each day reached the high nineties, except for Pablo's third day out, when the thermometer spiked at just over 103 degrees. These were temperatures in the shade. But where Pablo went, shade was rare.

By August 18, McGee was losing hope. By August 20, five days after Pablo went into the desert with no more than two days' worth of water, McGee went back to the routine of his camp. Pablo had either found a way out or was dead.

Early in the morning on August 23, McGee heard a sound coming up the otherwise silent desert gorge in which he camped. It was a low groan carried on the dry desert air. The groan came from Pablo. More accurately, in McGee's words, the groan came from "the wreck of Pablo," what was left of Pablo after eight days in the desert.

McGee described what he found in his 1906 paper "Desert Thirst as Disease":

> Pablo was stark naked; his formerly full-muscled legs and arms were shrunken and scrawny; his ribs ridged out like those of a starving horse; his habitually plethoric abdomen was drawn in almost against his vertebral column; his lips had disappeared as if amputated, leaving low edges of blackened tissue; his teeth and gums projected like those of a skinned animal, but the flesh was black and dry as a hank of jerky; his nose was withered and shrunken to half its length; the nostril-lining showing black; his eyes were set in a winkless stare, with surrounding skin so contracted as to expose the conjunctiva, itself black as the gums; his face was dark as a Negro, and his skin generally turned a ghastly purplish yet ashen gray, with great livid blotches and streaks.... His extremities were cold as the surrounding air; no pulsation could be detected at wrists, and there was apparently little if any circulation beyond the knees and elbows; the heartbeat was slow, irregular, fluttering, and almost ceasing in the longer intervals between the stertorous breathings.

Pablo's unwinking eyes were blind. Speech was out of the question. He had long since lost the ability to swallow. McGee poured water onto the skin of Pablo's face, chest, and abdomen. The skin shed the water at first and then absorbed it, like a dry sponge. McGee rubbed water into Pablo's nearly dead extremities. After a half hour Pablo could swallow. After an hour he could drink. Within two hours he could eat. At three hours—just after sunrise—he could, with help from McGee, walk the remaining short distance into camp. That evening, Pablo urinated.

On his last day out, Pablo had crawled seven miles over stones and cactus thorns. The journey left him with cuts and bruises that had swollen as he moved slowly through the desert. His hands, wrists, feet, and ankles had swollen, too. Now his breathing was hoarse and sometimes spasmodic, a blending of hiccups and retching that wracked Pablo's entire body and made him vomit. McGee gave him bismuth and pepsin-pancreatin tablets and simple camp food. After two days, Pablo's bowels were working. After three days, his vision and voice returned. He seemed to reawaken to the world. He recognized McGee and McGee's camp assistant. But the spasms kept coming. McGee feared for Pablo's life.

On August 27, a man named Jim Tucker, along with several friends of Pablo, showed up in camp with a four-horse wagon. They had heard that Pablo was lost and had come out to find and recover his remains. Instead they found Pablo himself, not quite dead. McGee, knowing that Pablo would die if he stayed there in the camp, urged the men to take Pablo to Wellton. "I judged," wrote McGee, "there was an equal chance of getting the patient alive to Wellton."

The patient lived. By August 31, he was "deliberately and methodically devouring watermelons." Within a week, he was gaining weight, coming in at around 135 pounds. He was cheerful, according to McGee, who reported only two permanent effects from the adventure: Pablo Valencia, the former sailor who knew firsthand of raving thirst, of desert thirst as disease, had lost most of his hair, and what remained had turned iron gray.

※

My companion and I wander the deserts near Las Vegas by car and by foot. They are higher than Death Valley and therefore cooler, cool enough to support cacti, wet enough for shrubs. They are cooler, but not at all cool.

At Red Rock Canyon, we walk through an abandoned sandstone quarry where a steam-powered truck hauled rock to the railroad in the late 1800s. The heat for the steam was generated by oil burners. The water was hauled in or caught in arroyos during the short rainy season or carried from a natural rain trap in the rocks nearby, a place called the water tanks.

We walk past a roasting pit, a shallow depression used by the Southern Paiute and Mojave people and maybe earlier tribes to heat blocks of stone that were then used to cook meat. In summer, it is hard to imagine cooking, but in winter temperatures drop below freezing. Now we walk again under the full light of the sun, over sand and rock and gravel through open country with scattered low-growing plants.

McGee, seven years before he found the wreck of Pablo Valencia, described desert plants. "The only plants able to survive the desert heat and drought," he wrote, "are water-storing monstrosities, living reservoirs like cacti and agaves." Small scattered yuccas and branching Joshua trees occasionally stand above the surrounding shrubs. There are barrel cacti, now favorites with landscapers. There are desert trumpets, *Eriogonum inflatum,* with swollen stems that were once used as pipes for smoking Indian tobacco. Creosote bushes are spaced almost evenly across the desert floor, as if part of an orchard, each bush sending up a dozen small trunklets, the branches of these trunklets holding dark green resinous leaflets in winglike pairs. Their spacing—the spacing that gives the desert a sense of order, a sense of openness—comes from the ability of each plant to suck every last drop of water from the soil reached by its shallow, outward-running roots.

A little higher, where the flat desert floor begins to rise, there are mesquite trees with leaves that resemble those of peas and taproots that drill down as far as 150 feet to find water. Mesquite seedpods can be dried and ground into a coarse flour called pinole, possibly the flour that Pablo Valencia and Jesus Rios carried into the desert.

We walk upward between rock walls that frame a corridor with a floor of sand as fine as sugar. We find a ponderosa pine, gnarled and hard bitten by wind and heat and lack of water and the rocky ground on which it struggles. The pines are remnants of earlier forests that thrived here in a cooler time, all but disappearing near the end of the Pleistocene, leaving only a few straggling colonies hanging on in shaded canyons, struggling to survive in a warmer world.

We find what at first glance I mistake for saltbush, its leaves narrow and thick, suggesting an *Atriplex* of some kind, but with a second glance I see that I am wrong. I see acorns. It is a canyon oak, adapted to the desert but as much an oak tree as the great shade tree live oaks of Florida and Georgia, as much an oak as the red oaks used for ship planks and the white oaks used as hanging trees, all species of the genus *Quercus*. And coming over a rise and downward into a sand flat, I see what look like willows, plants that go by the common name "desert willow" but are not willows at all, not even in the willow family, the Salicaceae. They are *Chilopsis linearis*, of the family Bignoniaceae. The Spanish common name is accurate: *mimbre*, meaning "willowlike," because their leaves are long and narrow, like those of willows, and their trunks are tall and slender and seldom straight, like those of willows. And, like willows, they love water. Somewhere below the ground here, not too far down, the dusty ground must be damp.

We move farther downslope, losing a few feet of elevation, and break out of the willowlike thicket to stand in a quarter acre of rushes, two feet tall, some dried and brown, others a lush green that speaks of recent water. The ground itself is dry. We are standing in the dust of a bone-dry marsh. A crust of algae separates my boots from the loose sand floor of the canyon.

I pick up a handful of the crust. Intact, it might make excellent tinder, but between my fingers, with the slightest grinding motion, it turns to dust.

This is the sort of place where a man like Pablo Valencia might desperately dig, looking for water. If lucky, he might find moist sand, something to put in his mouth. If he were extremely lucky, he might find enough water to wet a bandana, and he could squeeze the water from there into his mouth and then hold the damp cloth between his lips, sucking in an extra hour of life.

I measure the temperature of rock surfaces. Beneath plants, shaded by leaves and branches, the rocks and sand are cooler, at eighty-nine degrees. Between plants, bare rocks exposed to the sun come in at 138 degrees. But here is an oddity: the tops of plants, fully exposed to radiant heat from the sun, baked to the same extent as the rocks on the desert floor, do not exceed 105 degrees. They reflect light from the sun but also cool themselves by giving up water. Inside their leaves and stems, water molecules dance wildly with the energy of heat. A pore opens— a stomata, whose main purpose is to allow carbon dioxide into the plant. When the pore opens, water is lost, taking with it the heat that made the water molecules dance. The same physics cool our bodies when we sweat. The water leaves the plant or the person, turning to vapor and taking with it a measure of heat.

When a person sweats, if the sweat does not dry, it does not cool. In a Louisiana salt marsh, with humidity over 90 percent, where mildew outruns evaporation, sweating is of little use. Here in the desert, sweat dries. Sweat dries quickly, pulling heat away, but also sucking away water needed for survival.

I point my infrared thermometer at my companion's forehead. She comes in at ninety-one degrees, normal for skin temperature.

For the plants, as for Pablo Valencia, life is about heat and water. The plants that cannot survive the open heat or the scarcity of water live in shaded canyons and at higher elevations and near water traps. Some of the desert plants that grow only near water—the water spenders, as they are sometimes called—have to deal with the salts that accumulate near desert springs. Some would be at home in salt marshes. There is pickleweed, *Allenrolfea occidentalis,* also known as iodine bush, a plant that looks like the glasswort of the Gulf of Mexico coast but grows taller, forming dense shrubs. Its green succulent stems taste like salt. They are, in fact, full of salt. When a stem becomes so salty that it can no longer survive, the stem dies and is shed, and the plant itself survives.

Another strategy, useful for plants out in the open, away from easy water: grow quickly after rain, shed seeds, and wait for the next rain. Or leaf out following rain, then shed leaves after exhausting the rainwater to stand dormant with bare stems, like the thorny ocotillo, *Fouquieria splendens,* which, live and wet, stands as an upright green tassel but, live and dry, loses its leaves and turns brown and thorny and looks dead. Another strategy: curl leaves during the dry season and uncurl them during the wet. Or develop leaves with thick cuticles that hold water like waxed bags. Or alter the leaf pores, the stomata, by hiding them in pits or by growing fringing hairs around them, in both cases protecting them from dry winds. Or reduce the size of leaves, leaving less surface area through which to bleed moisture. Or, like the cactus, convert leaves to spines and relegate photosynthesis to the stem.

Some adaptations cannot be seen. Water is always needed to convert light and carbon dioxide to carbohydrates, the energy molecules of life. But how the light and carbon dioxide and water interact varies. Most plants—even desert plants like the creosote bush—follow a common path, a path that demands a constant supply of water, a path that demands open pores throughout the day to supply carbon dioxide. Without water, plant pores close, and photosynthesis stops. With water, a three-carbon carbohydrate forms.

Some plants—corn and saltbush—vary this theme, leaving their pores open during the day but making more efficient use of water, demanding high light and only photosynthesizing in inner cells, forming a four-carbon carbohydrate. And a few plants—succulents and cacti and some orchids—open their pores only at night, when the air cools. In the cool darkness, they pick up carbon dioxide through the pores and convert it to an acid. During the day, the pores close, but the reaction used to convert carbon dioxide to an acid reverses, releasing the carbon dioxide within the plant, allowing it to interact with water and light to form carbohydrates.

William McGee died in 1912, but if he were around today, I would write a fan letter, thanking him for "Desert Thirst as Disease," his story of Pablo Valencia. An article like his, I would tell him, changes the way one sees the world. It opens one's eyes. But, I would tell him, desert plants are not water-storing monstrosities. They are not any one thing at all. They are many plants, with many tricks, thriving in conditions that at first glance appear insufferable and deadly.

※

A half billion years ago, a sea washed through this part of Nevada and California. It was a time before dinosaurs, before warm blood, a time when backbones debuted in jawless fish, a time when plants were struggling for a foothold on land. It was a time of a distinctly different geography.

A few hundred million years pass, and a desert replaces the sea, a desert hotter and drier than the deserts we know today. The wind lifts red sand to form Sahara-like dunes. Time passes, and the climate changes again. Water moves through, calcifying the dunes, forming sandstone. Two hundred million years come and go. Four thousand years ago, the Gypsum People show up, inhabiting this place at a wetter time, a cooler time. Another two thousand years pass, and the Basketmaker People arrive, sandal wearers, contemporaries of Iron Age villagers in western Europe, contemporaries of Christ, contemporaries of Roman emperors. Flash forward again to nine hundred years ago and the Southern Paiutes show up, nomadic people moving with the changing seasons of increasingly dry country, builders of roasting pits. Flash forward to now, to today. The climate change experts say that these deserts will grow warmer and drier. Marginal habitat will become less habitable.

In our wanderings, my companion and I see rock art. Most of it is graffiti of obscure meaning, straight lines and squiggly curves and crescents, but also stick figures of people and animals and even rounded-out figures of animals that could be fat goats or bighorn sheep. Some of the rock art may have been left by Basketmakers. Some is younger, the work of Southern Paiute. It is light on dark, made by scraping away the blackened surface of the rock, leaving behind an image like that of a black-

and-white negative, but without any sense of three dimensions. If the taggers who drew here meant to leave a message, that message remains encoded, unexplained, maybe nothing more than a record of their passing, homage to the animals and the earth and the plants and the sky.

An antelope squirrel darts across the rocks, frantic sudden movement in a timeworn landscape. It scales a nearly vertical cliff, running across brilliantly lit rock art that glares in the sun.

I shoot a temperature of the rock art: 144 degrees.

In the 1930s, a University of California biology professor squatted in the desert watching lizards. The lizards were leashed in a manner that allowed them to move toward shade but not to reach it.

The professor discovered the obvious. He discovered that lizards died in the heat. "Here was I," he wrote, "a heat-generating animal with a naked, unprotected skin, surviving even longer exposure while dozens of reptiles were killed in minutes by overheating." He discovered that the smaller animals died quickly, that it takes less time to roast a small turkey than a large turkey. "Some of the smallest reptiles," he wrote, "died within sixty seconds after being scooped out of their underground shelters into the blazing sun of the surface."

He also discovered that the animals lost their ability to move well before they died. Self-rescue became impossible. He discovered too that their tolerance of cold far exceeded their tolerance of heat. They could cool thirty degrees below their optimal body temperature without loss of their abilities, but warm them more than ten degrees above their optimal body temperature, and they would suffer, in his words, "crippling effects."

All of this, it turns out, is also true for humans.

Lizards survive the heat by avoiding it, hiding in burrows or under bushes. Humans sweat. For humans, life at high temperatures returns again and again to sweating. Humans evolved in the tropics, a place that was hot but where water was abundant. We sweat to shed heat. We sweat to expose water to air, and that water evaporates, taking with it unwanted and potentially deadly heat. Adult humans have something like three million sweat glands. Individuals who grow up in extreme heat have more sweat glands than those who grow up in cold climates, but three million is a typical number.

Perspiration separates humans from dogs and cats and pigs, which shed heat by panting. It separates us, too, from birds, also panters. In terms of dumping heat, we primates stand closer to the other sweaters, to donkeys, camels, and horses.

We are warm-blooded. By virtue of breathing, of a beating heart, of firing neurons, we generate our own heat. *La respiration est donc une combustion.* At rest, we generate enough energy to power a hundred-watt lightbulb. Moving, contracting and relaxing muscles, we convert the chemical energy of food into the work of motion, but most of the energy—more than three-quarters of it—becomes heat. Moving, we generate enough heat to power ten lightbulbs. The chemical reactions in our cells produce enough energy to power a well-lit room, and just as in a well-lit room, most of that energy is lost to heat.

In the desert, the air and the sun and the ground add heat. Stand still and feel convection currents in the air moving heat across your skin. Stand in the sun and feel radiant heat. Sit on a rock and feel conductive heat.

Fail to dump heat, and the body's temperature rises. Add six degrees in core temperature, and suffer heatstroke. The skin goes dry, coordination diminishes, the mind becomes delirious, the body convulses with epileptic seizures. Self-rescue becomes impossible. Coma follows. Without help, only death remains.

❋

"As far as I am aware," wrote the medical doctor W. Hale White in 1891, "no explanations have been offered of the mode in which in the process of evolution cold-blooded animals became warm-blooded."

The difference between the cold-blooded and the warm-blooded is not as clear as Doctor White suspected. Take fish: cold-blooded, with temperatures matching those of the surrounding water. Unless the fish you take is the swordfish, whose innards include musclelike tissues that serve no purpose other than warming the blood supply to the brain and eyes. Or unless you take bluefin tuna, whose constant swimming generates heat, and whose blood vessels are arranged in such a way that the heat is conserved, with warm blood coming from the muscles donating heat to cold blood coming from the gills.

Body temperature of a swordfish, around the brain and eyes: as much as twenty degrees higher than that of the surrounding water.

Body temperature of a bluefin tuna: as much as twenty-five degrees higher than that of the surrounding water.

Then we have the camel, with a body temperature that drifts between the low 90s and 110 degrees, its thermostat adjusting to save water, to limit the need to sweat. Or certain bats, with temperatures when active at just under ninety-nine degrees, about the same as humans, but with temperatures dropping to those of the surrounding air when resting, and dropping close to freezing during hibernation. Or the arctic ground squirrel, with a body temperature when active similar to that of humans, but during hibernation dropping into the high twenties, below the freezing point of fresh water, not warm-blooded at all when curled in winter burrows. The camel, certain bats, the ground squirrel: warm-blooded, sort of, if one does not look too closely.

The earliest mammals pumped warm blood. They appeared two hundred million years ago, denizens of the late Triassic, a time when the supercontinent of Pangaea was just beginning to break apart. They probably came from cynodont therapsids, beasts that in appearance looked half lizard and half dachshund, egg laying but probably with hair, more warm-blooded than cold.

Warm blood flowed through the arteries of birds, too, appearing in the Jurassic, fifty million years after the early mammals gained a foothold. The birds probably descended from the theropod dinosaurs, which themselves may have sported warm blood.

Body temperature of the once living but forever extinct cynodont therapsids and theropods: unknown and unknowable.

⁂

Heatstroke kills hundreds of people every year. In the hot summer of 1980, thousands died. The heat wave of 1988 took thousands more.

Among weather-related events that cause death in the United States, heat waves outdo hurricanes, tornadoes, and floods. The elderly and the young are most susceptible. City dwellers, surrounded by concrete, are more susceptible than rural residents. In a typical year, seven hundred Americans die in the heat.

The numbers will go up. The number of elderly people increases every year, the number of city dwellers increases, and the temperature rises. Air-conditioning, en masse, triggers blackouts. Without air-conditioning, building temperatures skyrocket.

Researchers pull numbers from computer models. They predict that the average number of heat-related deaths in America will increase from about seven hundred per year to more than three thousand per year in the next four decades.

✳

Pablo Valencia did not experience heatstroke. He avoided the sun, walking at dawn and dusk and at night. He hunkered down when temperatures pushed ninety degrees. And he dumped heat by sweating, his meager water supply evaporating on his skin, taking heat with it. In this way, Pablo Valencia did not allow his body to reach the temperatures at which heatstroke occurs, at which systemic failure rips through the body, spiking the core temperature and killing the brain.

But he could not avoid dehydration and heat exhaustion.

Sweating or not, water is lost. Water moves through the lungs to be expelled with exhaled air. It moves through the nose. Water moves through the skin. Insensible water loss—water lost not as sweat but as vapor seeping through the skin—continues even in death. In the desert, a human corpse dries into a mummy within days. Pablo Valencia, when he collapsed near McGee's camp, was alive but in the early stages of mummification.

Sweating differs from insensible water loss. The body warms, and the hypothalamus triggers hidrosis. Put another way, the hypothalamus triggers sweating. Sweating dumps water through pores. We notice sweat. We sense it. We feel its wetness, and we feel the cooling of evaporation.

A well-hydrated 155-pound man—a man the size of Pablo Valencia before he went into the desert—can be thought of as a semiporous animated membrane containing forty-two quarts of water. About half that water is inside cells, and the other half is between the cells. About three quarts of the man's forty-two-quart total is blood plasma.

In the desert, it is possible to dump more than a pint of sweat in a single hour. Stay long enough to dump a little more than nine quarts and face shock.

After exhausting his canteen, Pablo Valencia, by the time he collapsed near McGee's camp, had lost well over two quarts of water. He drank his own urine until it became very bad, "mucho malo," as he reported it. He chewed on insects. He sucked water from plants. He found a green scorpion, ground off its stinger with a rock, and dined on its fluid-filled carcass. And all of this kept him alive long enough to leave him groaning on the ground, sending a hoarse moaning call through parched desert air and up the canyon to McGee's campsite, to rescue, to water.

Sweat contains salt. Dump heat by losing water, and dump salt. A person accustomed to the desert—a person like Pablo Valencia—produces sweat with as much as twelve times less salt than a person new to the desert, but the person still loses salt. K. N. Moss, working with miners in 1922, was the first to link hot conditions to the symptoms that come from an imbalance in the body's salts. The heat in which the miners labored was the heat of the earth, the heat found in the lower reaches of deep coal mines. The men worked short days at the coal face—only five or six hours—but during those short days they drank little, believing that too much water in hot conditions was bad for them, that they could suffer from water poisoning. They ignored their thirst and in exchange accepted fatigue, disorientation, and muscle cramps. Moss called the cramps "miner's cramps."

In the heat, with little water, with the body's salt levels out of balance, heat exhaustion becomes inevitable. The symptoms: fatigue, headache, pale skin, clammy skin, mild fever, nausea and vomiting, cramps, dizziness, fainting. And thirst.

During his first day without water, Pablo Valencia may have been talking to himself, babbling about ice cream or a stream from a childhood memory or a canteen. By August 20, 1905, three days before reaching McGee's campsite, Pablo's strength failed. His fatigue was so severe that he would sit for a while and then crawl. His vision blurred. The mountains and the creosote bushes and the mesquite jumped around, moving back and forth, as if seen through a layer of water. His eyelids by this time were stiff, and the tip of his tongue was hard enough to feel odd against his teeth. His face and lips dried into an unnatural smile, a monstrous fixed grin beneath glazed and staring eyes. At times, Pablo found his way with his hands, feeling the trail as he crawled. He believed that his partner, Jesus Rios, had abandoned him in the desert. Pablo kept going, motivated, he later said, by the desire to knife Jesus Rios. Pablo's lips dried further and split open and for a time oozed thickened blood that dried and resplit, eventually leaving his lips curled outward. His gums, too, dried and bled, for a moment giving a sensation of moisture in the mouth.

Unable to see or think clearly, Pablo crawled past a guidepost marking the direction to a spring. When he tried to walk, he fell, and getting up, he fell again. His tongue, by now entirely stiff and badly swollen, extended past his dried and curled lips into the open air. His eyelids, previously stiff, were now cracked. Buzzards landed next to him, within reach. He imagined, in vivid hallucinatory detail, his own death.

McGee saved Pablo and then used him as a case study. McGee described the phases of desert thirst as disease. Pablo was very close to McGee's final phase. "In this final phase," he wrote, "there is no alleviation, no relief save the end; for it is the ghastly yet possibly painless phase of living death, in which senses cease and men die from without inward—as dies the desert shrub whose twigs and branches wither and blow away long before the bole and root yield vitality."

Had Pablo Valencia wandered in Death Valley instead of the higher and cooler desert near Yuma, he would have died.

William McGee ended his paper on Pablo with a note about

Death Valley. On August 31, 1905, on the same day that Pablo was "deliberately and methodically devouring watermelons," McGee read a press dispatch announcing a thirst-related fatality in Death Valley. It was the thirty-fifth thirst-related fatality in Death Valley that year.

※

My companion and I move north, onto the Nevada Test Site, a desert weapons-testing ground the size of Rhode Island. A group of forty-niners passed this way, taking a southern route to California and a supposed shortcut that turned deadly. They later became known as the Death Valley Forty-niner Party, but at this point they were only starting their desert hardships. "We had been without water for twenty-four hours," wrote one of them, "when suddenly there broke into view to the south a splendid sheet of water, which all of us believed was Owen's Lake. As we hurried towards it the vision faded, and near midnight we halted on the rim of a basin of mud, with a shallow pool of brine." That faded lake, that mirage, were it real, would sit here in the middle of the Nevada Test Site.

Today the test site is secure, closed to casual wanderers, to forty-niners. We are here under the care of a guide. Our guide spent his career testing nuclear weapons. For a time he lived here, within the test site, in the government town of Mercury, Nevada. He claims to have once witnessed the creation of a fireball three miles wide. He is retired now, uninterested in ethics debates, dead certain that powerful bombs and the threat of mutual destruction saved the world. He is short, talkative, amicable, a wearer of cowboy boots, a bantam rooster with thumbs hooked into his belt. He is gray haired and wiry and dried out from years in the desert, years immersed in government bureaucracy, but he is as energetic as an antelope squirrel.

We pass simple wooden benches lined up in rows, bleachers

in the desert. Our guide talks of watching tests from these bleachers. He talks of a nuclear artillery round that was shot six miles from the muzzle of a cannon, landing in the distance with a fifteen-kiloton explosion—an explosion equivalent to that of fifteen thousand tons of TNT. He shows us pens where pigs, monkeys, and cows were caged at increasing distances from ground zero. He claims that he once barbecued a survivor. He shows us the remains of bomb shelters, domes of concrete blown open to expose metal reinforcement. We stop at the remains of a railroad trestle, its I beams sheared.

Trees were stuck into the ground to form artificial forests. Slow-motion photography developed for bomb tests shows the trees steaming as the radiation hit, then igniting. Then the shock wave came, extinguishing the flames like a child blowing out birthday candles, and the tops of the trees bent hard away from the blast, then back toward the blast, back and forth, reverberating with the shock wave.

We visit the remains of a doomsday town, a mockup of suburban America, where cars were parked and mannequins were stationed doing what suburban Americans did in the 1950s, washing dishes and playing in the yard and carrying briefcases. With unabashed delight, our guide says that one of the workers assigned to the creation of a doomsday town positioned two mannequins, naked, man on top, in a second-story bedroom.

The mannequins would have seen a flash of light. They would have felt sudden heat. Those far enough removed to survive the heat would have seen dust racing toward them across the desert floor. When the blast struck, it would have felt at first like a strong wind but would have increased suddenly, sweeping entire houses away, killing mannequins by the score. Farther out, cars parked broadside to the blast tumbled sideways, rolling across the desert. Cars parked toward the blast survived the shock wave, but the heat blistered their paint.

Our guide talks of sitting five miles from a test, where he felt the shock wave and the extreme heat. The shock wave passed, bounced off the mountains behind him, and returned. He crouched in a trench as the shock wave and its hot wind passed overhead. He was sunburned.

After several hundred nuclear bomb detonations, in 1963 a treaty ended above-ground testing. The government learned how to dig. Dig a shaft, lower a bomb into the shaft, plug the opening, and set off the bomb. Eight hundred and twenty-eight bombs were set off below ground. Each shot was named: the Uncle, the Bandicoot, the Gerbil, the Stones, the Pleasant, the Ticking.

Now we are sixty-five miles from Las Vegas, in Yucca Flat, the most bomb-blasted and irradiated piece of real estate on earth. We drive past crater after crater, some in groups, others standing alone on the desert floor.

In Area 10, we stop to look at the Sedan Crater. On July 6, 1962, the government chose to test a thermonuclear bomb—a hydrogen bomb—in a shaft 636 feet deep. Sedan, exploding, generated temperatures around twenty million degrees. In contrast, the surface of the sun, at a mere ten thousand degrees, would seem air-conditioned.

Sedan lifted a dome of desert earth thirty stories into the air. Beneath the ground, the earth was vaporized, leaving a cavity of heated gases. The gases cooled and contracted, lowering the pressure within the cavity. The ceiling of the cavity collapsed, chimneying, and the desert fell inward. The crater is as unnatural as rock art. At one-fifth of a mile across and more than three hundred feet deep, it is the ultimate graffiti.

Four years after the explosion, fifteen pounds of Sedan's radioactive soil were boxed up, shipped to Alaska, and spread on the ground at a place called Ogotoruk Valley, above the Arctic Circle, thirty-two miles from the tiny village of Point Hope, at a location where the federal government once planned to dig a harbor using hydrogen bombs. Upon hearing this, Ogotoruk Valley becomes another place I must visit.

※

In Hiroshima, half a world away and more than a half century ago, survivors of the world's first atom bomb attack fell ill. Most survivors who had been within a half mile of ground zero, near the Gokoku Shrine, died within hours or days of the blast. At first doctors thought that they were dying from burns or from injuries caused by the shock wave. The doctors were moving quickly from one victim to the next, treating burns and lacerations and broken bones in what had become, in an instant, a ruined city. But very quickly a pattern emerged: nausea, headache, diarrhea, malaise, massive hair loss, and fever. What is now called acute radiation syndrome (ARS) was killing off survivors. Among other things, the radiation attacked blood cells as they were generated in bone marrow. A few days later, when these damaged blood cells entered the bloodstream, survivors fell ill.

At Hiroshima, two key symptoms became apparent: high fever and low white blood cell counts. Patients with fevers that remained high or with white blood cell counts that dropped below one thousand generally died. Fevers could climb as high as 106 degrees.

※

In Old English, fever was *fefor,* said to have come from the Latin word *febris,* which itself may have come from the Latin word *fovere,* meaning "to warm" or "to heat," as in warming a kettle of tea or heating a bowl of soup. *Fovere* can be traced to an earlier word, something like *dhegh,* from the hypothetical proto-Indo-European language of nearly six thousand years ago, where it may have meant "burn" and may have led in other directions to mean "heat" and then "day," recognizing that daylight brought warmth. There is also a possible link to the Sanskrit *bhur,* meaning "to be restless." In German, it is *fieber,* and in Swiss it is *feber.*

Ancient Egyptians, Chinese, and Mesopotamians had dissected enough bodies to understand anatomy, but fever was attributed to evil spirits.

Pliny the Elder, Roman soldier and scholar, completed his thirty-seven volume *Natural History* in AD 77, filling it with observations and anecdotes, covering topics from astrology to agriculture. He wrote of fever: "Some persons are distressed by a perpetual fever. Such was the case with C. Mæcenas; during the last three years of his life, he could never get a single moment's sleep. Antipater of Sidon, the poet, was attacked with fever every year, and that only on his birthday; he died of it at an advanced age." Pliny wrote that all men experienced fever, with the exception of one particularly healthy specimen, Xenophilus the musician, who lived to be 105. Pliny knew about malaria and recognized its pattern of high temperatures coming and going, cycling through as if driven by a clock.

Years after writing about fever, Pliny died during the eruption of Vesuvius.

In the middle of the nineteenth century, in the time of Faraday, the German physician Carl Wunderlich grew interested in fevers. Over sixteen years, he compiled data on twenty-five thousand patients. He used a thermometer that could require twenty minutes to stabilize. Wunderlich promoted the reality of fever as a symptom of illness rather than an illness in its own right. In 1871 he wrote, "All abnormal temperatures denote a disease, but all diseases do not show an abnormal temperature." From another passage: "The temperature may be determined with a nicety which is common to few other phenomena. The temperature can neither be feigned nor falsified. We may conclude the presence of some disturbance in the economy from the mere fact of altered temperature."

Wunderlich was not interested in why warm-bloodedness and certain sicknesses were accompanied by the fever response. "And though theoretical questions as to human temperature and kindred subjects must not be overlooked," he wrote, "my purpose has been to prepare from these notes a practical book."

Fever arises when a foreign substance, such as a lipopolysaccharide in the cell wall of a bacterium, triggers the release of cytokines from white blood cells. The cytokines signal a part of the brain just above the brainstem and straight back from the bridge of the nose—a part of the brain called the hypothalamus—to turn up the heat. Muscles tense, the body shivers, and veins constrict near the body's surface. The body generates more than the normal amount of heat and sheds less. The white blood cells' response to bacteria resets the body's thermostat. The set point moves from just under 99 degrees to 101 or 102 degrees. Or dangerously higher.

The fever, under some circumstances, helps the patient. White blood cells proliferate when the body is warm. As a consequence, certain disease-causing microbes suffer. Under other circumstances, fever is one more discomfort along the short road to death. At 103 degrees the patient begins to lose the normal sense of self and starts to drift into a haze of fading self-awareness. The patient's set point—the normal temperature of the warm-blooded creature—has been reset, the internal thermostat turned up. The patient, forehead burning, cannot feel warm.

Between 1348 and 1350, one-third of Europe's population succumbed to bubonic plague, with fever and swelling in the armpits, groin, and neck. Another sickness known as the Sweat, perhaps a form of hantavirus, struck England in 1485, 1508, 1517, and 1528. It left patients at first very cold, then suddenly hot and sweating, and then often dead. In the 1520s, smallpox brought by Cortés's thugs overwhelmed the Aztec empire, bumping up their thermostats, killing millions, leaving behind stories of the dead abandoned in their homes. Near the end of the eighteenth century, yellow fever took about one in twenty New Yorkers: fever and chills, internal bleeding, headache, backache, slow heartbeat, vomiting, jaundice, all followed by a short intermission and apparent recovery, but in the final act a return of the symptoms and ultimately death. Deadly influenza visited the world in 1732 and 1733, in 1761, in 1775 and 1776, in 1847 and 1848, in 1850 and 1851, and again and again, each time bringing fever and death, especially in naive populations, those without previous experience of the influenza virus, populations like those of Hawaii and Alaska.

From an Alaskan narrative written during the Great Sickness of 1900, which seems to have been a strain of influenza accompanied by measles and smallpox that came into the region with gold prospectors and miners:

You enter a tent and you see a man and his wife and three or four children and some infants lying on a mat, all half naked, coughing up bile with blood, moaning, vomiting, passing blood with stools and urine, with purulent eruptions from the eyes and nose, covered with oily and dirty rags, all helpless, and wet and damp day and night.

From the famous Alaskan missionary doctor Joseph H. Romig:

They were cold, they were hungry and thirsty and weak, with no one to wait on them. The dead often remained for days in the same tent with the living, and in many cases they were never removed.... Children cried for food, and no one was able to give it to them. At one place some passing strangers heard the crying of children, and upon examination found only some children left with both parents dead in the tent.

※

In my Death Valley hotel room, I listen to the air conditioner and read about Kuda Box. Born in Kashmir in 1905, he became a firewalker at age fourteen. At thirty, he walked for Harry Price, a well-known psychic researcher in England. Price called the walks "experiments."

In the experiments, Box walked across a fire pit that measured eleven feet long, six feet wide, and nine inches deep. The fire, made from "some seven tons of oak logs," as well as charcoal, ten gallons of paraffin, and "fifty copies of the *Times*," was ignited at 8:20 in the morning. Box walked at three in the afternoon, when the surface temperature in the fire pit was 800 degrees. The temperature within the bed of coals was higher, reportedly at 2,500 degrees. Kuda Box walked through twice.

"He was quite unharmed," Price wrote.

The experiment did not end there. "Some amateurs who attempted to duplicate the feat were burned," Price reported, "but not severely." Price could not explain why Kuda Box could walk through unharmed, while the amateurs, presumably volunteers, could not.

"Firewalking," he wrote, "was in no sense a trick." He thought it might be an act of faith or the result of short contact with the hot coals or "that there was a knack in walking."

If it was an act of faith, it was a faith I envied. If it was the physics of short contact time, so be it. If it was a knack, it was a knack I could develop.

I find a video clip about Tolly Burkan, another firewalker. Burkan was born in New York City in 1948 and later moved to California. Since the 1970s, he has promoted the benefits of firewalking: confidence building, spiritual renewal, healing, team building, something to break the tedium of day-to-day life.

In the video clip, Burkan talks to students of the Firewalking Institute of Research and Education. "The next time you're in a situation that used to intimidate you," Burkan tells them, "you will remember, 'I walked on fire, and if I can do that, certainly I can go in there and ask for a raise!'"

Burkan, slender, balding, wearing a tie, talks to the camera from a stage of sandy ground in a clearing surrounded by scrubby trees. "When you are in the right state of mind," he says, "the blood flows through the soles of your feet and takes the temperature away from the tissue."

A close-up shows Burkan's smile. The smile involves not just his mouth and teeth but his cheeks and eyes and even his ears, which pull back and flatten as the smile stretches across his face. The smile is almost goofy. It is certainly disarming. It is the smile of a man who would be impossible to dislike.

Burkan's clients stand in the soft twilight around a fire pit similar to the one that Kuda Box walked through eight decades earlier. The clients clap rhythmically—two claps slowly, a pause, then three quick claps. They chant the word "Yes!" The chanting matches the beat of the clapping, the first two yeses delivered slowly, the last three in rapid succession: "Yes!—Yes! Yes!-Yes!-Yes!"

Burkan is the first through the fire. He is barefoot. Like his students, Burkan chants, "Yes!—Yes! Yes!-Yes!-Yes!" He walks through the fire, his hands stretched upward, the gesture of a supplicant. His pace is ginger, but certainly not panicked. His slacks—business casual—are rolled up at midcalf, well clear of the fire.

His students, one by one, follow.

"It was pretty hot," one young woman tells the camera.

"Whether you're a physicist and you believe in these laws of physics," Burkan says, "or whether you're someone who just believes in me because you trust me, as soon as you walk into the fire with a belief that you are not going to burn your feet, you are in a different physiological state than the person who thinks they're going to get burned."

Burkan believes there are three million firewalkers in the world and three thousand firewalking instructors on six continents.

Another Burkan quote: "I've seen people horrifically burned."

Tolly Burkan's enthusiasm convinces me. I must become one of the three million. I need to walk through fire. I telephone the Firewalking Institute of Research and Education, the organization that Tolly Burkan started. I talk to an instructor. He tells me that Burkan is retired, not available.

"I'm a writer," I tell him, "and I want to give it a go."

The institute is in the business of walking corporate types through fire as a team-building exercise, but it also offers instructor training. "Become a Firewalking Institute of Research and Education Certified Firewalking Instructor," reads the advertisement, "and learn so much more than firewalking."

"I may want to become an instructor," I say.

"Things in life," the instructor tells me, "are a lot easier than they seem to be."

Sometimes the instructor uses bonfires. Other times he uses railroad fires, long corridors of flame through which his students parade barefoot.

"Firewalking," he tells me, "is one of those things you don't think you can do."

I do not argue this point. Instead I ask about injuries.

"There are some," he answers. "Usually no more than hot

spots and blisters like you might get from a long walk in new shoes. It's a mind-set. Tell yourself you'll get hurt, and you will."

The temperature of the hot coals through which a firewalker walks, he tells me, is between 700 and 1,500 degrees.

From my 1963 edition of the *Encyclopedia Britannica*: "The interesting part of firewalking is the alleged immunity of the performers from burns. On this point authorities and eyewitnesses differ greatly."

Through the telephone I hear a truck engine. I hear highway noise. "I'm on my way to Malibu," he tells me. In the truck, he carries firewood. After Malibu, he will teach a course in Dallas, and then in Spain, and then another in Dallas.

It is hard to talk over the highway noise. I ask if he ever teaches in the desert.

"We teach wherever people want to learn," he tells me. "But in the desert we have to be careful about fire restrictions."

He is thinking of deserts wetter than Death Valley, about deserts with abundant fuel, about dry chaparral, about places where outside fires require a burn permit, places where there is a lot more to burn.

※

Our first walk in Death Valley threw my companion and me off, pushed our electrolytes out of balance, leaving us worn out, on the edge of debilitation, two Alaskans sick with heat, feeling slightly feverish. We do not suffer from miner's cramps, but even now, two days after our first walk, we feel the sense of disorientation that miners once described. Sweat, in the unacclimated walker, is as salty as the blood and plasma and other fluids. It is likely that we lost close to a quart of sweat during our four-hour walk, and with it we lost salt. Upon returning, we drank water. We drank water to a fault. We pushed ourselves toward hyponatremia, toward water intoxication. We pushed ourselves toward a sodium imbalance associated with nausea, headaches, confusion, lethargy, fatigue, appetite loss, vomiting, convulsions, and comas.

The mechanisms of hyponatremia are complex and confusing. Fluid levels within cells change and fluid levels between cells change. The plasma becomes salt poor. The kidneys grow confused.

It takes time for the body to regain balance. We drink enough to be hydrated, but we pay for our walk through the desert. The water has to find its way through the body, into cells and between cells. Salts have to be redistributed.

Acclimatization is possible, but not in a way that conserves water. The opposite happens. Humans adapt to heat by sweating more, not less. After days and weeks in the heat, sweat becomes more dilute. The acclimated body conserves salt. Blood plasma levels increase, and with that increase comes an increase in performance, an apparent tolerance to the heat. More importantly, there are behavioral adaptations. People learn to move slowly. They learn to stay in the shade. They learn to orient their bodies so as to expose as little surface as possible to the sun. They learn not to hike in open deserts in the early afternoon with only a gallon of water to share.

My companion and I adapt by heading to cooler elevations, into the mountains above the desert to a place where we can look down into Death Valley. We drive to the Panamint Range, stopping at an elevation of seven thousand feet. Here we find trees, junipers with fresh pale berrylike cones and pinyon pines with spreading shade-tree branches, cones the size of baseballs, and needles pointing upward as if in constant prayer. There is Mormon tea, *Ephedra cutleri*, known for its medicinal properties. There is cactus, too, twelve-inch-tall Mojave prickly pear with inch-long gray spines. Chickadees are busy in the juniper branches, and the buzz of hummingbird wings cuts the air.

Next to the road, we walk along a line of twelve kilns built by Chinese laborers in 1877 to turn trees into charcoal. The kilns stand in a neat row. They are dome shaped, made from brick, each kiln maybe twenty-five feet tall and thirty feet across. A single kiln could be loaded with forty-two cords of wood. The wood would smolder for a week and then cool for a week, leaving behind two thousand bushels of charcoal. The charcoal was light enough to be hauled to a mine in the treeless valley below, where it was used to smelt lead and silver.

I enter the kilns one at a time. I smell smoke from fires that burned more than a hundred years ago.

We head uphill, moving slowly, worn out and weak from our days of desert walks. As we move higher, the trees grow more scattered. The forest takes on the appearance of a savanna. Here and there, between living trees, old stumps stand out, monuments to the kilns below.

In two and a half hours, at nine thousand feet, we reach Wild Rose Peak. The summit, windswept, is more or less treeless. We sit on a rock and look over Death Valley in silence. The salt flats on the floor of the valley appear to be flooded. The uninitiated, looking down on this valley, might believe that it contains a pleasant shallow lake. We drink from our water bottles. It is comfortably cool here, even in the sun. With altitude, temperature drops. A rule of thumb—seldom correct, but often close—puts the temperature drop at about four degrees for every thousand feet of elevation. While our summit just touches 80, the valley below bakes at 116 degrees.

Another day passes, and we drive to Death Valley's Ubehebe Crater, a half mile across and six hundred feet deep, wider and deeper than the Sedan Crater, but natural and less radioactive. It may have formed around six thousand years ago, but estimates vary by thousands of years. Better known is how it formed: hot magma rose up from the depths of the earth, encountered ground water, and turned the water to steam. The steam, expanding, threw out shattered rock, sand, and ash as far as six miles.

We hike to the bottom. The crater is less conical than Sedan, with one side rising almost vertically and the other sloping steeply downward, an incline of loose sand and crunchy gravel that fills our shoes.

At the bottom, desert trumpet and desert holly grow, along with scattered creosote bushes, some ten feet tall and casting long morning shadows. Leaf-cutter ants have made small craters of their own, eight inches across. They move out from their own craters, their nests, and march across the Ubehebe Crater floor, bringing bits of leaves back and toting them down into their tunnels. The tunnels can be twenty feet deep, stuffed with scraps of leaves on which the ants cultivate a fungus. The ants eat the fungus.

Climbing out of the crater, we see other wildlife. There is a coyote, as handsome as a groomed dog, said to smell water from miles away. In an erosion gully, we see a Gila monster, a foot long and squat. Later, near the crater rim, we see a rodent, a kind of rat, maybe a pack rat. Certain rats, adapted to the desert, produce urine five times as salty as seawater—they could drink seawater and then filter out the salt to survive, peeing what Pablo Valencia would call "mucho malo."

The human kidneys, unlike the kidneys of certain rats, cannot extract freshwater from seawater. For humans, to drink seawater is to die.

In July 1945, the USS *Indianapolis* delivered the bomb that would be dropped on Hiroshima. Four days later, the *Indianapolis* was torpedoed. It sank in minutes, leaving nine hundred men floating in life rafts and life jackets. Four days passed before a plane spotted survivors.

From sailor Woody James: "The next morning the sun come up and warmed things up and then it got unbearably hot so you start praying for the sun to go down so you can cool off again."

The men grew thirsty. "Some of the guys been drinking salt water by now, and they were going berserk," James recalled. "They'd tell you big stories about the *Indianapolis* is not sunk, it's just right there under the surface. I was just down there and had a drink of water out of the drinking fountain and the Gee-dunk is still open. The Geedunk being the commissary where you buy ice cream, cigarettes, candy, what have you. 'It's still open,' they'd tell you. 'Come on, we'll go get a drink of water,' and then three or four guys would believe this story and go with them."

Three hundred and seventeen men—one-third of those who had survived the sinking itself—were rescued. Of these, about a hundred had survived those four days without freshwater. Of these hundred, none had drunk seawater. A medical officer reported that those who drank seawater became sick, delirious, and combative. They drifted away from the group or swam away, apparently maddened. No one who drank seawater survived.

In the morning, my urine flows weak and deep yellow. My lips are cracked, and my throat is scratchy in a way that drinking water does not soothe. My muscles ache. Despite the heat, I feel strangely chilled, as if I am fighting the flu. There is no question that my set point has been reset, nudged upward to a slight fever. I am thirsty yet have the feeling that a drink of water is just too much trouble.

My companion has trouble with her earrings. Her ears have swollen, closing the piercings.

We drive through the desert, stopping here and there for short walks. We drink water.

Late in the afternoon, we visit Badwater Basin. At 282 feet below sea level, it is Death Valley's lowest point.

A hot, hard wind, a moisture-stripping wind, sweeps down from the mountains, carrying air that is more or less trapped in the valley. The hottest air rises from the surface of these salt flats, and in rising loses pressure, and in losing pressure cools. Before it can reach the tops of the mountains, before it can escape this valley, it grows cool enough to sink, and in sinking gains pressure, and in gaining pressure warms. It steals further warmth from the exposed earth of the mountainside.

The temperature in the shade at Badwater Basin is 124 degrees, but the only shade is that of my own shadow. The ground is almost pure salt and almost pure white. My metal pencil feels as hot as a stovepipe.

In Badwater Basin, I envy camels. I envy the way in which their body temperature fluctuates. I envy their ability to sustain a body temperature of 108 degrees without sweating. The camel's loose coat reflects the sun. The surface of its coat may be thirty degrees warmer than its skin. Its head is plumbed in such a way that incoming blood is cooled, to protect the brain. When exhaling, it traps moisture in its nose. It produces urine as thick as syrup and twice as salty as seawater. It produces dung so dry that it can be burned fresh from the rump. It can store twenty gallons of water in its stomach and intestines. When dehydrated, it can gulp down fifteen gallons. Within a day, the badly dehydrated camel, given enough to drink, is rehydrated, healthy, and ready to go.

I am not a camel. Humans sweat with core temperatures as low as 100 degrees and just under half are dead from heatstroke before the core temperature reaches 108 degrees.

Twenty thousand years ago, Badwater Basin would have been six hundred feet underwater. The water flowed down from glaciers in the surrounding mountains, and when the glaciers disappeared, the flow stopped. The lake dried. The salts from the lake accumulated here. The salts form sharp ridges two inches tall that outline irregular polygons two and three feet across, stretching into the distance, stretching across this reflective basin, arid in the extreme.

The wind grows in strength until it impedes walking. It sweeps past us like the wind from a suddenly opened oven door. It is like the wind described by our guide at the Nevada Test Site, the wind from a bomb test that he felt passing overhead as he crouched in a trench. It engulfs me, blasting heat across my skin, between my fingers, behind my ears. It dries my eyes and my nostrils.

Over the mountains, thunderclouds form. I face into the wind and watch them approach. My shadow leans out across the salt. I smell ozone. I hear thunder. The clouds move quickly. A rainbow appears above Death Valley. I can see the rain, but it does not reach the ground. It disappears as it falls. But then I feel a drop, and another, and another. For twenty seconds, fine beads of water hit the ground, evaporating instantly on the hot salt. And then the rain is gone.

Chapter 2

UNMANAGED FIRE

Idrive to Santa Barbara to visit a photographer. After his neighborhood burned, the photographer and his wife, wearing
gloves and dust masks, sifted through the ashes. They invited
friends. In what had once been their home, they found charred
and broken dishes, globs of metal that had once been spoons
and forks and clocks and lamps, burned books with pages
turned to fragile ash. They found a copy of *Br'er Rabbit* burned
around the edges, and they saw it as an object of novel beauty,
an object of fire art. A magazine page, burned but for the image
of a man's face, emerged from the ashes. A mosaic made by the
photographer's mother surfaced, charred and in places melted,
but worth salvaging. Half of a coil of green garden hose survived, its other half blackened and melted.

Four days after the fire, a friend found a wineglass in the
ashes, picked it up, and was burned through her gloves.

The photographer, wearing his dust mask, in jeans and a

white T-shirt, posed for a picture next to a metal file cabinet. The cabinet stood charred and warped, slumped under its own weight, a file cabinet that Salvador Dalí might have painted. The file cabinet once held the photographer's best work: striking portraits of plantation workers in Sierra Leone, asbestos miners in Russia, a tribal woman smoking a pipe in Burma, Afghani men crouched in conversation. Before the fire, it had been a file cabinet full of images. After, it was a slumping metal hulk full of ashes.

The photographer talks quietly, forming thoughtful sentences that often end with an upward inflection. I ask how the loss of his house and his possessions and his photographs affected him.

"It was a sense of relief," he tells me, "a feeling that we could make a new start."

He and his wife and their friends gathered bits and pieces of fire-ruined goods. He talked to neighbors and borrowed artifacts from homes that had burned to ashes. In his studio, he put the artifacts in a light box. With soft light shining from below, the fire artifacts were framed in angelic white. With more light from above, the objects themselves were brilliantly lit. They are cataloged by the street addresses of now gone homes. Many are abstract shapes.

One picture shows what look like the metal remains of an antique revolver from 1325 West Mountain Drive, the wooden parts gone, the iron rusted. From 45 West Mountain Drive: a lightbulb that has lost its shape, its glass heat-softened on one side, allowing it to lean over with a dent in its heat-softened head. From the same house and the same fire: a ceiling fan with its blades and wires burned away, and a heart-shaped Christmas tree ornament now lopsided, a sheet metal angel of the sort meant to sit on top of a Christmas tree but now tragically drooping. From 350 East Mountain Drive: a pile of coins, copper faces green and twisted and unrecognizable. And from

245 East Mountain Drive: what looks like it could once have been a doll or a figurine, seated, with stubby feet and hands and thick limbs, hairless, its skin melted into globs and whatever face it had almost gone, its mouth now a gaping round hole that could be moaning or screaming, its eyes and nose burned flat.

❋

The earth has been around for four and a half billion years. Fires such as the one that burned the photographer's house did not exist for the first four billion of these years. The fire triangle—the triangle of heat, oxygen, and fuel—was incomplete on two sides. There were lightning strikes and lava flows to provide heat, but no oxygen and no fuel. Oxygen began to accumulate two billion years ago, pumped out of the oceans by algae, with levels becoming comparable to those of today five hundred million years ago. Around then, in the time of trilobites and ammonites, plants crept ashore. On mudflats and rock shelves close to the tide line, mats of algae developed. Moss evolved. A sort of fungus or lichen grew twenty feet tall with a trunk three feet thick. Club mosses and horsetails and ferns appeared on the scene.

Another name for these plants: fuel. The primitive land plants closed the fire triangle. The most ancient charcoal in the fossil record appears in the early Devonian, four hundred million years ago.

The powdery yellow spores of club mosses would eventually be used in fireworks and explosives. In the 1800s, along with candles, alcohol, phosphorus, and hydrogen, Faraday, during his candle lectures, burned club moss spores, calling the spores by the scientific name of the plant. "Here is a powder which is very combustible," he wrote, "consisting, as you see, of separate little particles. It is called *lycopodium,* and each of these particles can produce a vapor, and produce its own flame; but, to see them burning, you would imagine it was all one flame."

As plants crowded the land, they pulled carbon dioxide from the air and then died, locking that carbon dioxide away. The net effect was a cooling climate, a reverse greenhouse effect eventually offset as fuel accumulated and wildfires became common.

✻

The nineteenth century French mathematician and physicist Jean Baptiste Joseph Fourier, known for his work on vibrations and heat transfer, realized that the earth, based on its distance from the sun, should be a much colder planet. Its average temperature should be close to freezing, but in reality it hovers around sixty degrees. Why? He considered that heat might come from space, "from the common temperature of the planetary spaces." He considered that the "earth preserves in its interior a part of that primitive heat which it had at the time of the first formation of the planets." But in the end he focused on the atmosphere and oceans. For the most part, his work focused on the distribution of heat. "The presence of the atmosphere and the waters," he wrote, "has the general effect of rendering the distribution of heat more uniform." Heat from the tropics found its way north and south. But somehow—and he did not know how—the earth's atmosphere trapped the heat of the sun. By virtue of the atmosphere, most of the earth was habitable.

Fourier, thanks to papers he published in 1824 and 1827, is widely credited with the discovery of what would become known as the greenhouse effect.

✻

Before the photographer's house burned, before it could burn, the atmosphere evolved. Four and a half billion years of change set the stage for the Tea Fire—the fire that took a photographer's house and photographs, took his neighbor's house too, burned or partly burned 219 homes spread across three square miles, converting prized possessions into misshapen lumps and fire art, but miraculously killing no one in its flames.

Heat, the first side of the fire triangle, was provided by college students who lit a bonfire up above Mountain Drive, near the abandoned ruins called the Tea House. The students said the fire was out before they left, but authorities suggested that the bonfire still held life, warm coals that the students missed. Fuel, the second side of the fire triangle, stood as dry chaparral, a shrubby mix of chamise and scrub oak and California lilac and black sage, with a few Coulter pines and the remarkably flammable Australian eucalyptus with their crowns well above the chaparral, perhaps even drier than normal with the effects of climate change. Oxygen, the third side of the triangle, rode in the air on a wind known locally as a sundowner, a wind that comes tumbling down the Santa Ynez Mountains, running toward the sea, warming as it loses altitude and gains pressure, like the winds that blow downward to fill Death Valley. Its warmth stripped the remaining moisture from plants while driving their flammable vapors into the air. The wind fanned the fire, turning warm coals hot, then hotter. Flames appeared around dinnertime and, pushed by the sundowner, moved downhill toward the houses.

As these things go, it was an average fire, a historical footnote, one fire in many. It was contained and controlled within days. During that time, more than five thousand homes were evacuated. Seven hundred and fifty-six firefighters were mobilized, with sixty-two fire engines. Something like six million dollars was spent in three days, and a thick-limbed doll was melted into an objet d'art, heated by the Tea Fire, along with a lightbulb, a pile of coins, an antique pistol, and a copy of *Br'er Rabbit*.

✳

I drive a road through the hills and along the ridgeline above Santa Barbara with a scientist who studies, among other things, wildfire fuel. He looks at photographs taken from satellites and airplanes. From these photographs, he makes maps. Meaning in no way any disrespect to his science, I think of him as a mapper.

We can see Santa Cruz and the three islets of Anacapa Island and oil platforms off the coast, and the city of Santa Barbara surrounded by suburbs, and the suburbs surrounded on three sides by fuel in the form of dense chaparral. In places, orchards and highways form what may be thought of as firebreaks, separating the fuel of chaparral from the fuel of suburbia.

The Spaniards who ranched in California called the dense shrubby vegetation chaparral because it reminded them of scrub oak hillsides in Spain called *chaparro*. Cowboys chasing cattle into the chaparral invented chaps to protect their legs. In the heat of summer, they would have seen the chaparral burn just as it burns today.

The mapper carries binoculars and a camera and several maps. "Avocado and citrus orchards and golf courses make good firebreaks," he tells me.

He talks on and off about climate change. He wonders if the chaparral is entering a time when the fire season will extend throughout the whole year, when fires will burn in winter as they burn in summer. He comments, too, on the spread of housing and human activity deeper into the chaparral. It may be that an increase in fire frequency will come from climate change, but it could just as well come from increased sources of ignition, from the close proximity of humans, matches, and fuel.

The road, narrow and steep and winding, will not let two cars pass comfortably. It is not the sort of road I would associate with southern California. It is a road that would be at ease in parts of Montana or Alaska.

We pull off to look at fuel. "This one," the mapper says, "is *Ceanothus megacarpus,* or big-pod ceanothus." It is a lilac, in the buckthorn family. "Only the leaves and small branches burn in the first fire," he tells me. "The leaves burn off and the trunks stay behind. But the trunks die and dry out and are burned in the next fire."

Big-pod ceanothus is not the sort of plant one would grow in a firebreak. Its leaves carry flammable resin. Its seeds drop to the ground beneath it, saturating the soil, but only germinate after a fire. They germinate in response to heat or smoke or the presence of charcoal. A stand of big-pod ceanothus is a sure sign of returning fire.

"They take about eight years to mature," the mapper says. "If the fire return rate is too quick, they will be replaced by laurel sumac. That's what happened in the Santa Monica Mountains."

If the land does not burn, the big-pod ceanothus grows old and eventually dies off, to be replaced by other plants. But, in fact, the land always burns. Big-pod ceanothus does not worry about old age.

At another stop, we look at spiny ceanothus, *Ceanothus spinosus.* It is wetter here, along the edge of a gully. It occurs to me that a wetter area would be a good place to avoid fire, to hide, to hunker down. I have heard that deer and bears often head into gullies during fires. But, in fact, the gullies burn hard and fast. Fire rips through gullies like smoke up a stovepipe, so much so that firefighters sometimes call the gullies chimneys.

There is jack pine, too, and ponderosa pine, and eucalyptus, each adapted to fire but in different ways. The cones of the jack pine only open in response to fire, and they open slowly, after the fire has passed, dropping their seeds onto the naked soil left behind by a hot burn. The seeds survive temperatures of one thousand degrees. The ponderosa pine survives fire by virtue of its thick bark. The eucalyptus, imported to California from Australia, full of flammable sap, reputed to explode if sufficiently preheated and ignited, possesses epicormic buds—growth sites hidden beneath the protective cover of bark, ready to sprout if the outer skin of the tree is damaged by fire.

Climate experts expect most of the United States to become hotter and drier. They expect less rain in summer, and less snowpack in winter.

The mountain pine beetle and the spruce beetle attack and kill conifers. The spruce budworm eats their needles. The beetles and the budworms are heat limited. Cold weather kills them, and warm weather speeds growth and breeding. As the climate warms, the beetles and budworms become more abundant.

In a warming world, they attack and kill entire forests, leaving standing dead wood over thousands of acres. On the one hand, this means an abundance of dry firewood, leaving some climate experts expecting more fires. On the other hand, when pine and spruce die, their needles disappear. Without needles, there is less fuel. Some climate experts expect fewer fires.

The chaparral, with its occasional Coulter pine or eucalyptus, with its dense shrubs covered by tiny leaves full of resinous oils and its accumulation of dead branches and twigs, is the most flammable plant community in North America. Chaparral fires have been compared to gasoline fires, with entire hillsides igniting in an instant to become a sheet of flame blasted by wind and whipping along the ground and reaching three and four and five times higher than the height of the shrubs themselves.

The mapper and I drive farther along the ridge and stop again near La Cumbre Peak. Fifteen months after the Tea Fire, the Jesusita Fire tore through here, taking an area ten times the size of New York's Central Park. Eighteen thousand people were evacuated. Cal Fire, the largest fire department in California, posted the following message: "Wildland brush fire driven by slope, erratic winds and single digit humidity's are causing significant runs with extreme fire behavior."

A Santa Barbara County Fire Department captain reported that the fire was "moving very, very rapidly." He called it what it obviously was: an uncontrolled wildfire.

Flames overran a Ventura fire engine and its five-man crew. Plastic on the inside of their fire engine melted. Afterward the men were evacuated with first-, second-, and third-degree burns.

Before it was over, flames caught more firefighters. The Jesusita fire injured thirty firefighters but, amazingly, killed none.

Within two weeks the Jesusita fire was completely contained. This does not mean the fire was out but that it was surrounded, subdued, and to some degree under control. The small task of mopping up remained. The fire burned across thousands of acres, but the first hot flush crossing the landscape left behind unburned fuel. That unburned fuel smoldered and flamed erratically, in small patches, smoke coming off the ground over here, charcoal working its way up the inside of a tree trunk over there, and just over there an unburned shrub dried and heated enough to suddenly transition into combustion.

In some settings, fires smolder for days, eating through the duff that lies on the ground. In other settings, fires can smolder for years. For the purposes of mop-up, chaparral is forgiving in comparison to, for example, stands of spruce and pine. Mop-up is easier in chaparral because the fire takes most of the fuel and leaves behind exposed mineral soil, but that is not to say that it is easy. Crews are tired. Smoke and ash fill the air and saturate clothes and skin and hair. The heat of the ground finds its way through boots.

The crews search for smoke. They scan the ground with infrared scopes, looking for hot spots. When they find something smoldering, they attack with water or hand tools. In places, they work through the goop of fire suppressants dropped from airplanes and helicopters.

The Jesusita Fire burned in part over regrowth that had sprung up after the Painted Cave Fire, which had burned twenty years before. The Jesusita Fire was contained in part by the more or less barren ground left by the Gap Fire, which had burned eighteen months before. Older firefighters on the Jesusita Fire told younger firefighters about fighting the Painted Cave Fire, and younger firefighters told rookies about fighting the Gap Fire. The firefighters could, without leaving the greater urban area of Santa Barbara, also spin tales about the Refugio Fire, the Polo Fire, the Coyote Fire, the Romero Canyon Fire, the Sycamore Canyon Fire, and the Eagle Fire. Cumulatively these fires burned an area twelve times the size of Manhattan.

From our vantage points along the road, the mapper points out a wide swath of charcoaled hillsides. Through binoculars, the blackened skeletons of trees stand above a ridgeline with exposed tan bedrock and scattered boulders. I cannot tell where one fire started and another ended. The mapper finds a subtle change in color, from one shade of green to another, barely perceptible to my eyes, and says that it marks a fire line from an earlier fire.

He points to a stand of dead trees with their branches and twigs intact. "Sometimes fire will pass an area without burning it," he tells me, "but the trees are killed by the heat or by the ash."

Today there is no fire. Smoke does not obscure the view.

Three paragliders hang in the air above us, and bicyclists pedal along the ridge road.

We scurry down a hillside of loose, sandy gravel. The smell of ash prevails.

In his work with aerial images of burned and unburned hillsides, the mapper looks at more than just color, and he maps more than just plant communities. "Water has a particular spectrum and signature," he says.

Moisture content goes into his computer models. He starts simply, looking at fuel and slope and moisture. From there, he adds wind. He tosses in subtle changes in fuel loading that let his virtual fire move in sudden leaps interspersed with stalls. This patch of ground might be beetle killed, that patch might have a stream running through it, another patch might have burned within the past few years and have little to offer in the way of fuel.

In computer models of wildland fires, a pixel changes from green to red, indicating ignition. That pixel ignites neighboring pixels. The fire spreads. Before long, the fire creates its own weather. The fire's own heat dries out moist vegetation. The fire's own winds roar up gullies. The fire reaches the crest of a hill, with an updraft on one side and a downdraft on the other. The fire leaps across three green pixels, leaving them untouched, to turn four more green pixels red.

The challenge is to come up with models coupling the reasonably static reality of the landscape with the overwhelmingly dynamic reality of an uncontrolled wildfire. The challenge is to create a model that captures land burning with a ferocity capable of creating sounds that firefighters have compared to the sounds of trucks, trains, tornadoes, hurricanes, and volcanoes.

The mapper tells me of measuring temperatures during fires. It is not unusual, he tells me, to measure temperatures hotter than two thousand degrees.

With the photographer, I drive into the hills behind Santa Barbara, toward where the photographer's house no longer exists. We stop at a neighbor's house. The neighbor is rebuilding. He has cut down the charred trees that once grew on his property and had them milled into floorboards for his new home. In the ceiling, framed in but not yet finished, he has installed a sprinkler system of the kind usually seen in public buildings. In a wall, he has framed in a fireplace. In his backyard, a few orange trees survive, standing amid burned stumps, their oranges shriveled and scorched and misshapen, oranges that would be at home in a heat-warped basket on top of a sagging file cabinet.

The neighbor points out the source of the fire, the point of ignition, near the Tea House. The Tea House is uphill from here. Despite its name, it was, before it was abandoned, before it burned, a tea garden. In the gardens stood a stone platform and three arches built in 1916. No longer used for tea parties, the overgrown gardens attracted kids with beer.

The fire swept downhill, driven by the sundowner wind. Witnesses would later describe the fire coming down the hill like a liquid, following low spots that funneled wind.

"I had maybe fifteen minutes of warning," the neighbor says. His family—his wife and daughter—were away. The warning came in the form of word of mouth. There were no police knocking at the door. No sirens sounded. He heard the noise of a hot wind and, before he drove away, the noise of the fire itself. He saw the flames.

"Flame is burning smoke," Aristotle wrote, "and smoke consists of air and earth."

It is hard to know if Aristotle understood the truth of his words. Flames glow as they do because fire releases fine particles of soot, and that soot, heated, glows, just as a toaster element glows from the heat of electricity, just as the filament of an incandescent lightbulb glows from the heat of electricity.

Faraday described glowing soot in his candle lectures. "You would hardly think that all those substances which fly about London, in the form of soots and blacks," he wrote, "are the very beauty and life of the flame. It is to this presence of solid particles in the candle flame that it owes its brilliancy."

Neither Aristotle nor Faraday was familiar with chaparral, but both men would have been quick to understand its nuances. When fire burns through chaparral, it goes in two stages: first comes the roaring flame of the crown fire, which passes quickly, and then the slower burn of the undergrowth.

"Fire may burn rapidly through the crowns of brush," wrote G. W. Craddock in 1929, "causing leaves and small stems to fall in flames to the ground where they ignite the litter to produce maximum surface soil temperatures in 2–9 minutes."

Two closely spaced points may suffer through different experiences. The first point might heat abruptly, spiking to 1,500 degrees in minutes, consuming all its fuel in an angry burst and then cooling quickly after the fire passes. The second point might be hit by a crown fire that heats the soil suddenly to 1,000 degrees before passing, after which it cools to 200 degrees before the ground fire arrives. Under the flames of the ground fire, the soil might reach 900 degrees. The ground fire might burn for two hours, working its way through leaf litter and duff and even burning the richest organic parts of the soil itself, ending with the softly glowing combustion of a few smoldering stumps and, five hours after the fire has passed, a soil temperature of 600 degrees. A day later, the ground, in places, might still burn through boots. The photographer and the neighbor discuss their memories of the fire. As they remember the Tea Fire taking their homes and their possessions and threatening their lives, I choose not to bring up Aristotle or Faraday or Craddock.

The neighbor's wife is a painter. Her paintings went the way of the photographer's photographs. After the fire, when the neighbor returned and searched through the ashes, he found lumps of gold that had once been his wife's jewelry. Fourteen-carat gold, typical of jewelry, melts at temperatures close to 1,500 degrees.

"I sometimes imagine the house burning room by room," the photographer says. "I think about the things that burned."

Afterward, he says, the first thing you buy is a toothbrush and pet food. You rebuild from there.

※

Ben Franklin, the man behind Philadelphia's first volunteer fire department, died in 1790, the year before Faraday was born. Franklin would not have thought of heat as an expression of molecular motion. In all likelihood, he thought of heat in terms of Lavoisier's caloric theory, as a subtle fluid.

Lavoisier and Franklin knew each other. In 1784 they served on a French commission that debunked the medical theories of Franz Mesmer. Mesmer believed that what he thought of as the vital force of animals—what he called their *magnétisme animal*—could cure illness. The essence of life, of health, flowed like caloric, as a subtle fluid. Mesmer believed that his subtle fluid could address stomach cramps and blindness and fevers.

Mesmer's techniques did not always work. Despite this, his name found its way into common usage. People can be mesmerized in any number of ways.

The commission on which Franklin and Lavoisier sat is often credited with developing controlled clinical trials, a technique used today to test drug therapies.

Also on the committee was Joseph-Ignace Guillotin. In 1794 the invention named after Guillotin ended the life of Lavoisier. "The Republic needs neither scientists nor chemists," ruled a

French judge, and Lavoisier lost his head, ending any opportunity for him to revise his caloric theory and to recognize heat for what it really is, an expression of molecular motion, not a fluid at all.

From the astronomer and mathematician Joseph-Louis Lagrange, soon after Lavoisier's execution: "It took only a moment to cause this head to fall and a hundred years will not suffice to produce its like."

❋

The photographer and I leave the neighbor's house to drive uphill to where the photographer once lived. Someone has cleaned up the rubble that was once his house. Someone has hauled away his melted file cabinet. At first it seems there is nothing left but a concrete slab and a beautiful round chimney made from rough bricks, tapering near its top, where the roof would have been. But odds and ends appear. We find an abalone shell, blackened and burned thin by the fire, but its mother-of-pearl inner face still intact. We find a battery, its label burned away and one end melted. The broken ends of wooden beams turned to charcoal jut from holes in the chimney, suggesting that there had once been a ceiling. On the ground, in splotches, are pools of hardened metal of the sort that the photographer had collected and photographed. In what was once the living-room floor, weeds poke up through cracked concrete next to a hardened pool of melted glass. There is a blackened barbecue grill. Next to it the plastic plumbing is still more or less intact, unmelted.

The photographer shows me where the kitchen was. He shows me where the bathroom was.

"Coming here after the fire," he says, "was like working through an archaeological dig, but what should have taken centuries took only a few minutes. It was hard to tell what things were."

We hear a dog barking. We hear the pounding of a hammer, of someone rebuilding. The landscape, though charred in places, is for the most part green again. Prickly pear cactus grows thick along a trail leading up from the remains of the house. Castor bean grows in dense thickets. Low weeds and grass grow on what may have been a patch of lawn.

The house next door did not burn. It sat in a pixel skipped by the flame. Other pixels, too, were skipped. There is, for example, an unburned tree. Another neighbor's house was gutted by flames, except for the bathroom, an unburned pixel, where towels hung on racks unscathed, and a plastic shower curtain hung unmelted. In the driveway of what had once been the photographer's house, the fire skipped past his wife's Alfa Romeo convertible. Everything around it was gone, but the car remained intact. Its vinyl roof did not melt. Its tires held air. It started. It ran. The photographer and his wife drove the car away, down the hill and away from the ashen neighborhood where they had lived, evidence that even the gods of fire love Alfa Romeos.

✹

I call the Firewalking Institute of Research and Education. I reach the instructor. "Ever been in a house fire?" I ask him. He has not. Nor has he been in a wildfire. But he knows firewalkers who live in the California hills, surrounded by fuel. He is not sure if any of them have lost their homes.

"Firefighters come in for our fire walks in Dallas," he tells me. "Some of them come on a regular basis. They tell people about skin melting. They tell them that the fire, the ash bed, will still be hot in the morning, maybe six hundred degrees. Their stories are great. It's important for everyone to be outside of their comfort zone. We want to stretch the students. We want them to understand that this really is hot. This could be dangerous."

❋

I meet a firefighter with twenty years of burning chaparral under his belt, and I ask how close he is to the flames when he fights fires on hills like these. It varies, he says, but he knows he is too close if his skin starts to blister.

We drive ninety miles north and east from Santa Barbara to park on a dirt trail inside the gate at Spanish Ranch. We leave the truck where a firefighting crew had set up a command post in 1979. It was from here that a team of now dead firefighters was sent up Sycamore Ridge, told to catch up with a bulldozer that was cutting a firebreak along the ridge. By then the fire seemed under control, almost contained, and the firebreak, running along the ridge, was well removed from what was left of the flames.

We walk uphill along Sycamore Ridge, toward where the men died.

This is scenic California ranch land with steep, rounded hills and sandstone bluffs above flat meadows, dry but not scorched. Brown and black and white cattle graze in the flat meadows, leaving nothing but nubs of grass where the fire started. On the hillsides, scraggly shrubs stand two or three feet tall, peppered across the slope and separated from one another by bare earth. Even now, during what passes for the wet season, dry chaparral paints the hillsides brown.

The chaparral would have been even drier after weeks without rain in August 1979. And the firefighting had started in the afternoon, after the sun flushed out what little water the plants had absorbed the night before. Despite this, it is hard to imagine that this place of stubby grass and sparse shrubs could carry a flame. It seems to me that the chaparral must have been thicker in 1979, that what we are seeing is incomplete regrowth even after thirty years.

The firefighter disagrees. "If anything," he says, "it is thicker now."

It is hard to imagine this place, even bone dry, burning with a ferocity capable of killing four firefighters.

Walking, we gain altitude quickly. Ten minutes above our truck, we stop at a marker. This is where firefighter Scott Cox, badly burned, emerged from the smoke.

We continue upward another ten minutes. Three stone cairns stand near where three bodies were found. Each cairn is adorned with a brand-new cloth badge from Cal Fire, reminding the world that these three men have not been forgotten, that Cal Fire remembers its fallen even after three decades have passed.

Around the cairns, thirty-year-old blackened nubs of burned-out shrubs remain, not quite hidden now by new growth, by new fuel.

By the time they reached this point, the firefighters would have been sweating from the climb. They might have noticed a wind suddenly against their faces, a wind that had not been blowing when they started uphill. This was more than a breeze. This was a wind of twenty-five or thirty miles per hour, the breath of a bellows exhaled into the base of the fire. They might have looked down toward the flat grazing land below, expecting to see a fire that was more or less under control. Instead they saw a conflagration.

The men, feeling the wind and seeing the flames, realized that their situation had suddenly gone from routine to potentially dangerous.

Somewhere ahead of them, the bulldozer was cutting trail, blading in the firebreak intended to surround this fire, to complete its containment. The men at this point were still thinking of catching up to the bulldozer, of completing their assignment. The machine was out of sight, behind the next hill.

Also out of sight was a tight gully. This gully became a chimney, sucking flames and heat upward. The fire erupted up the chimney with flames twenty-five feet tall. At the same time, the fire came across the hill, toward the men. The plants—first under the sun and now under the heat of the fire—had dumped natural vapors. Flames ripped through the vapors, flashing across the hillside.

As the men watched, within minutes, possibly within seconds, the potentially dangerous erupted into dangerous.

Ahead of the men, the bulldozer operator realized that the fire would overrun him and his machine. He took a few seconds to dig out a haven, four blades wide, pushing away dry shrubs, exposing bare earth, plowing away fuel. He parked his machine in the middle of this tiny haven. The air grew hotter and thick with smoke. As the smoke shifted and danced in the wind, a hot orange glow appeared and disappeared. The bulldozer operator lowered fireproof curtains over his windows. He wrapped himself in a fire blanket. And then, terrified, he let the fire burn across his machine.

Afterward, miraculously with only minor injuries, he described flames thirty feet tall burning across his position. The paint on his dozer had wrinkled. His radio antenna had melted.

The fire jumped the ridge without hesitation. Hot embers and burning leaves, caught by the afternoon wind, now supplemented by the fire's own wind, blew onto fresh fuel. Spot fires ignited. They grew with tremendous speed into a wall of flame. The crew, still making its way toward the bulldozer, was surrounded. They were contained by the fire. The dangerous had erupted into the deadly.

Later, from his hospital bed, Scott Cox described what had happened. The men trying to catch up with the bulldozer knew they were surrounded. Like the bulldozer operator, they knew that the fire would overrun their position. "We all lay down on the dozer line," Scott Cox said. "We all lay down on the trail."

They had decided to tough it out, hunkered down in their fire-proof clothes, lying flat against the bare earth of the dozer line.

"It got too hot, and it began to burn us," Cox said. "We got up again and ran."

Cox's three colleagues ran uphill. They probably died within seconds from inhaling superheated air, before their bodies were badly burned. Cox ran downhill. He covered his face with his hands and ran into the flames, hoping to break through to the other side, to where the fire had roared through and burned itself out, hoping to break through into a landscape of scorched but cooling earth, into the black. In doing so, he became a fire artifact, burned over 60 percent of his body. His rescuers described him as bent over like "ancient man," the classic posture of the badly burned. They cooled his body with water and wrapped him in a burn bag. He looked up and said, "I'm gonna make it."

My guide shows me a photograph of Cox's Nomex shirt. Nomex is flame resistant. Exposed to extreme flames, it absorbs heat and swells, closing off spaces in the weave and further insulating its wearer. The Nomex shirt in the photograph is charred. It is a fire artifact, but one that lacks anything resembling aesthetic beauty.

Scott Cox, in the hospital, told a doctor, "I didn't think Nomex would burn."

Sixteen weeks after the fire, Cox's colleagues were shaken by his appearance. The athlete they had known, the twenty-six-year-old man who had stood six feet, two inches tall and weighed 220 pounds, who had played college football, who was more than capable of fighting wildland fires day after day, was gone. "It was a shocking experience," one of the men said of the visit.

Cox died three months later.

✳

With the firefighter, I stand in a parking lot next to a green fire engine similar to the one that was burned over in the Jesusita Fire. It is, in fact, the firefighter's engine, the engine with which he fights fires.

It is packed with gear: an organized rack of brass fittings, hoses of different sizes and lengths, special packs that ride low on the hips for wildland firefighters who spend most of their days bent over and hacking away at the ground, other packs built to hold hoses that uncoil behind firefighters as they walk or run from the engine toward the fire. There are special wrenches for joining one piece of hose to another. There are drip torches for lighting backfires, the controlled fires set by firefighters to burn out a piece of ground, to destroy fuel before the main fire arrives. There is an assortment of hand tools—shovels and hoes and a specialized rake called a McLeod and the most famous of all tools for wildland firefighters, the pulaski.

The pulaski is the wildland firefighter's tool of choice, half ax, half mattock or pick, the ideal hand tool for cutting a firebreak or for mopping up. The Forest Service owns thousands of pulaskies, and it may be said that no wildland fire engine rolls without at least one pulaski on board. It was invented by Ed Pulaski, a forest ranger in Idaho at the time of the nation's greatest fire, the Great Fire of 1910, also called the Big Blowup or the Big Burn. An area about the size of Connecticut burned in Idaho, Montana, and Washington, torching three million acres, killing at least eighty-seven people. Pulaski—the man, not the tool—led his crew of forty-five ill-trained firefighters through a fire that was long past the point of unpredictability.

During the Great Fire of 1910, winds exceeded fifty miles per hour and carried with them ash and embers and smoke. Sap boiled in trees just before the trees ignited. Creek water warmed up and steamed and flowed hot and full of dead fish. Columns of what looked like smoke and steam occasionally flashed into flame.

During that fire, ranger Ed Pulaski led his shaken men into an abandoned mining tunnel. To control his half-panicked crew, he drew his revolver. He made the men lie facedown, like one of Houdini's initiates in a burning cage. The fire sucked the air from the tunnel. To sit or to stand was to suffocate. Timbers shoring up the tunnel's entrance ignited. Pulaski himself eventually lost consciousness. His hair and face and body were burned, but he and most of his men survived.

His wife, seeing him after the fire, said that he staggered. His eyes were covered by bandages. His hair and hands were burned. His body was burned. In all likelihood, his lungs were burned. Breathing itself would have been a source of pain.

The government turned its back on the injured. Pulaski and other rangers used their salaries and savings to pay for medical treatment, both their own and that of some of the firefighters. And Pulaski, when he recovered, went into a blacksmith shop and made a pulaski. Among wildland firefighters, the story of Pulaski, the man, is almost as well known as the feel of the pulaski, the tool.

The four men who died at Spanish Ranch likely had among them a pulaski, but it can be said with certainty that they did not have fire shelters. Almost everything written about the fire suggests that they may have survived had they carried fire shelters on their belts. Now, because of what happened at Spanish Ranch, all wildland firefighters in California carry a fire shelter on the job.

My firefighter, my guide, pulls out a fire shelter. It is a tightly packed cube of folded foil and fabric, wrapped in plastic. The firefighter's first line of defense is to avoid being caught by fire, but these shelters have saved the lives of more than 250 firefighters who were surrounded and burned over.

The shelter, opened, is shaped more like an oversized sleeping bag than a tent. The fabric is aluminum glued to a fiberglass frame. The idea is simple: when surrounded and at risk of being overwhelmed by fire, find a firebreak, set up the shelter, crawl inside, and hope for the best.

The shelter will not survive direct contact with flames. Surrounded by fire but untouched by flame, the shelter reflects 95 percent of the fire's radiant heat. Still, the inside of the shelter will grow hotter than a dry sauna.

From the fire shelter manual: "When you are inside a fire shelter, breathe through your mouth, stay calm, and above all, stay in your shelter." To step outside, to panic and run, is to die.

The shelter, in direct contact with flame, fails. The conductive heat that comes from contact with the hot gas and glowing soot of the flame itself destroys the shelter. Flames in wildland fires average 1,600 degrees but can spike to temperatures much higher. The glue bonding the fiberglass flame to the aluminum foil breaks down at just less than 500 degrees, and the shelter delaminates. The aluminum melts at 1,200 degrees. The shelter can fill with smoke and fumes from glue and aluminum. The fumes can explode.

From a survivor: "The right side of my shelter delaminated, and the foil flipped over onto the left side. I really started to get burned at that point because the only thing that was on that side of my shelter was the glass mesh. There was still a tremendous amount of radiant heat coming off the surrounding area, a wind blew the shelter half back on to the other side, back to where it belonged, and it was like somebody closing a door on the oven."

I take the shelter from my firefighter and imagine myself in a fire, scrambling to find a clearing, somewhere without fuel, somewhere that will be beyond reach of the flames themselves. By now, I have ditched most of my gear, dumped it and run. If I am still thinking clearly, I have water with me, and my radio, and maybe my pulaski. Time allowing, I might use the pulaski to cut in a safe zone. Or, even better, I might find a bulldozer line or a gravel creek bed or a swath of burned-out forest or a broad ridgetop. I imagine smoke and heat and chaos and the knowledge that I may soon be dead. This close to the flames, strong winds will tug at the shelter. The noise of the fire will be disconcerting. The heat will burn my earlobes and blister exposed skin.

From the fire shelter manual: "The end facing the advancing fire will become the hottest part of the shelter. It will be easier to hold that end down with your feet than with your hands and elbows. Keeping your head away from the heat as long as possible will better protect your lungs and airways."

Firefighters are trained to deploy a shelter in less than twenty seconds. I shake mine open, use one foot to pin the shelter's lower wall to the ground, and pull the rest of the shelter over my torso and head, as if donning a hooded parka. I lie facedown on the ground, surrounded by the shelter. With my elbows and knees, I pin the shelter to the ground. I am following Houdini's orders, lying low in the cage, imitating Pulaski and his crew in the cave that saved their lives. I cup my hands around my face and breathe close to the ground.

From a survivor: "If you look at the burn injuries that I received, anything that was off the ground and certainly the things that were higher up in the shelter were the areas where I received the most significant burns."

From the fire shelter manual: "At a fire's peak, the noise will be deafening. You may be unable to hear anyone. Keep calm."

To the extent that it is possible, the manual encourages users to talk to their colleagues, to reassure one another.

From a survivor: "If you hear anything at all, the things you hear you don't want to hear, you wish you'd never heard."

Survivors report thinking that they would die in their shelters. They have compared the experience to that of being in a nuclear blast. They pray. They think of loved ones.

From the fire shelter manual: "No matter how bad it gets inside, it is much worse outside."

Typical burn-overs, typical entrapments, last fifteen to ninety minutes.

❋

Before the Spanish Ranch fire, Ed Marty, one of the men who died on Sycamore Ridge, made occasional public appearances dressed as Smokey the Bear, more properly and officially called Smokey Bear, an icon with a first and last name. "Only you can prevent forest fires," the bear said, on thousands of posters, echoing Gifford Pinchot, the first chief of the United States Forest Service, who once wrote, "Forest fires are preventable."

Pinchot was writing just after the Big Burn of 1910. The nation was focused on wildfires. Pinchot pleaded for funds to support fire towers. "It is a good thing for us to remember at this time that nearly all or quite all of the loss, suffering and death the fires have caused was wholly unnecessary," Pinchot wrote.

Pinchot also kept up a running dialogue with the spirit of his dead fiancée for decades, imagining her alive and well, in the room with him.

Smokey Bear came after Pinchot's reign. In 1942, just after Pearl Harbor, a Japanese submarine shelled the California coastline west of the Spanish Ranch Fire and north of Santa Barbara. Justifiably, people worried that shelling could ignite wildfires. The Forest Service established a wartime advertising program with posters and slogans: "Forest Fires Aid the Enemy" and "Our Carelessness, Their Secret Weapon." That same year, Disney came out with *Bambi*. Viewers saw Bambi smelling smoke. They saw Bambi flee from the flames. Bambi—the cartoon fawn, not the movie—was loaned to the Forest Service for one year to be used in a fire prevention campaign. When that year ended, the Forest Service had to switch cartoon characters. They came up with Smokey, originally naked but for his hat, eventually morphing into a beer-bellied, jeans-wearing bear, hat still intact, with suspenders and, quite often, a shovel.

Edward Abbey, author of *Desert Solitaire* and enduring symbol for the environmental movement, spent summers working in fire towers. Abbey, like most writers apparently starved for cash, found it convenient to let the government pay his expenses for the summer. "We are being paid a generous wage," he wrote, "to stay awake for at least eight hours a day." The wage was $3.25 an hour. Abbey sat in his tower, staring across the treetops, hoping to spot a fire, reading Robert Burton's *The Anatomy of Melancholy*, first published in 1621.

Abbey saw neither smoke nor flames. "If that idiot Smokey the Bear had his way," Abbey wrote, "all us firefighters would starve to death." And this: "When Smokey Bear says that only *You* can prevent forest fires, Smokey is speaking an untruth." Although Abbey was convinced that lightning ignites 90 percent of western forest fires, statistics lend at least some credibility to the idiot bear, showing that half or more of wildland fires are human caused.

But Abbey, a writer and activist, would have understood the bear's need for an ardent message. No iconic fire prevention bear could say something closer to the truth, like, "Only you can prevent *some* forest fires."

The bear exaggerated a bit but in so doing became, to all appearances, immortal. Unlike Abbey, the bear's name is protected by law. Also unlike Abbey, it is possible to buy a stuffed toy version of Smokey. A portion of the price of a stuffed Smokey goes to firefighting.

Ed Marty, before he died on Sycamore Ridge at Spanish Ranch, dressed up in his Smokey Bear suit and convinced parents and children that they should not toss cigarette butts out the car window or play with matches. Ed Marty was less beer-bellied than the cartoon bear, more barrel-chested, but his shoulders slumped forward slightly, and his eyes looked downward at the pavement. Wrapped in the bear suit, he appeared to be hot.

The fire that killed Ed Marty and Scott Cox and their colleagues, it turns out, was human caused. State employees were clearing grass along the road when a mower blade sparked against a rock. What was their purpose in clearing the grass along a remote stretch of two-lane highway? They were clearing grass to reduce the risk of fire.

※

When Scott Cox ran through the flames of the Spanish Ranch Fire, burning 60 percent of his body, he entered a new world, a world of pain in which he would live for the next 202 days. He was not alone in the world of fire-caused pain followed by death. In a typical year, more than three thousand Americans die in fires. For the most part, they are burned in house fires.

When people burn—when they are engulfed by heat and in contact with flames—they may or may not feel immediate pain. A person on fire is likely a person one step removed from panic or, more likely, entirely immersed in panic. The adrenalin load of a person on fire might kill a small dog.

There is also what has been called Wall's theory, or the gate control theory of pain. As a matter of coping, as a matter of survival, the perception of pain is modified by the brain. In a sense, the brain sends out signals saying, "This is too much," and the signals from melting skin and hair and hot gases making contact with the trachea are toned down. The brain controls the pain.

And there is this terrifying reality of the pain of burns: bad burns destroy nerve endplates. Heat melts the skin and dries it and the tissue beneath it. Then heat consumes flesh as if it were firewood. The receiving end of the sensory cells becomes fuel. Third-degree burns—what are increasingly called full thickness burns, burns that extend into the subcutaneous fat and muscle—take the top layer of skin and the lower layer of skin and the hair follicles and the nerve endplates, leaving behind charred meat. The fire takes with it the ability to sense pain at the very locations where the pain should be most severe. Instead the victim feels pain from the adjacent tissue, less burned and with nerve endplates intact.

Clothes and tissue continue to roast even after the fire is gone. Responders cut away hot clothing and try to cool burned flesh. The heat from clothes that have melted to the victim and the heat from the burned tissue move deeper into unburned tissue, into the body.

As the adrenalin wears off, the victim reacts to the pain. The pain comes not from the worst burns, where the neural endplates are gone, but from the surrounding areas. Morphine is administered, or pethidine, or fentanyl, or nalbuphine, along with tranquilizers.

From a New York City firefighter, after being trapped in a burning building: "Morphine does the job. You don't feel the pain. But I wouldn't recommend it to anybody. While I was taking the morphine those two weeks, I created a whole world inside my head. It was part reality and part hallucination. Some parts were terribly frightening. What was happening in my head wasn't pleasant at all."

The skin is gone. Without it, bodily fluids weep. A vein is found, and fluids are pumped in.

Skin, intact, acts to regulate the body's temperature. Skin, gone, cannot regulate the body's temperature. The victim, badly burned, shivers or is too hot.

Dead tissue is cut away. In places, it is scrubbed away.

The victim is wrapped in bandages. Bandages trap heat. The victim is moved to a temperature-controlled bed.

Over time, the neural endplates grow back. They sense the damage. They send pain signals to the brain.

Burn victims often experience hyperalgesia. Literally, to suffer hyperalgesia is to be in exceeding pain. But it is more than that. It is an increased sensitivity to pain. It is the sensation of pain in response to the merest touch, to the slightest vibration. It is the sensation of pain in the absence of an obvious cause of pain.

From Silas Weir Mitchell's 1872 book *Injuries of Nerves and Their Consequences*:

> Perhaps few persons who are not physicians can realize the influence which long-continued and unendurable pain may have on both body and mind. Under such torments the temper changes, the most amiable grow irritable, the bravest soldier becomes a coward, and the strongest man is scarcely less nervous than the most hysterical girl. Nothing can better illustrate the extent to which these statements may be true than the cases of burning pain, or, as I prefer to term it, *Causalgia,* the most terrible of all tortures which a nerve wound may inflict.

Bad burns affect more than just the tissue that has burned. They affect the organs. They have a systemic effect that must be overcome.

From a 1943 article in which three doctors summarized their experience with sixty-one cases of fatal burns: "This concerns not only the local skin and adjacent structures, but also such distant organs as lungs, lymph nodes, adrenals, kidneys, liver, and the gastro-intestinal tract." In the badly burned, for reasons that are not always clear, organs that were not damaged by the heat itself begin to fail.

The article includes a table with sixty-one entries. One column is labeled "Clinical Notes."

Clinical notes for entry number 1, a three-year-old: "Vomiting, convulsions, high temperature."

From entry 10, a four-year-old: "Peripheral circulation failed."

From entry 30, a nine-year-old: "Vomiting blood, delirious, coma, toxemia."

Commonalities emerge: twitching, convulsions, fever, and toxemia. Occasionally there is more detail. Entry 16 reports a temperature of 107 degrees, entry 18 reports a temperature of 108 degrees, and entry 48 reports a temperature of 106 degrees.

With entry 51, regarding a five-month-old baby, there is an inexplicable note: "Fried pancakes to chest."

The changes to the body introduce challenges to the administration of painkillers. In some patients, the metabolism of certain drugs changes as the body deals with its burns. Changes in blood flow change the way that drugs are delivered to the body. The liver may process certain drugs more slowly than it would in the unburned. The fire, gone now, affects the effectiveness of drugs in the badly burned.

In the privacy of my hotel room, I light my candle. I consider once again holding my palm to the flame, this time with the burned in mind, in an effort to better understand their experiences. But I cannot do it. To do it now would be to disrespect the pain of the truly burned. Ashamed, I extinguish the flame and spend thirty minutes trying to imagine the pain of the truly burned. I cannot, and for that I am grateful.

※

With the firefighter, I drive over two-lane highways surrounded by chaparral, surrounded by fuel. Clouds gather and thicken. Rain falls. We turn onto a dirt road and encounter a sign on a gate: "Area closed for public safety and watershed recovery due to damage from recent wildfires." The firefighter opens the gate, and we pass through onto the La Brea Fire site, burned six months before.

The truck slips in the mud as we move uphill. On a hilltop, we stand in the rain and look across a landscape that had been chaparral and will become, again, chaparral. Now it is bare earth and rock and scattered skeletons of shrubs set against a backdrop of rugged hills, almost as barren as Death Valley. But here and there new growth emerges. Specks of green show.

The view is more or less the same on both sides of the road. The fire jumped the road.

Ignition was somewhere out in those hills. It started with a cookstove at a camp used by marijuana growers. Something like ninety thousand acres burned, an area a hundred times the size of Central Park, ten times the size of Santa Barbara's Jesusita Fire. The response included 901 personnel on-site. Some would have been driving bulldozers. Others would have been swinging pulaskies. Still others would have been driving buses and filling out paperwork and operating radios and washing dishes.

In chaparral, plant resins leach into the soils. When fire heats the soils, the resins coat soil particles. After a fire, soils will be water resistant for a time, hydrophobic. After a fire like this, water runs off bare hillsides, picking up volume and momentum as it flows downhill, ripping out patches of ground as it goes. Mudslides follow chaparral fires. From where we stand, we can see two mudslides that, in populated areas, would have destroyed homes.

Certain insects come to fires. Most famous among these is

the fire beetle, *Melanophila acuminata,* known for detecting fires from miles away, flying in, and laying eggs in the dead but still hot skeletons of trees. Dead, the trees are defenseless. Burned trees do not exude resin to stop the beetle larvae as they feast on the dead wood.

To find fires, the beetles use receptors hidden in pits along the sides of their bodies. The heat warms tiny sacs in these pits, and the warmed sacs press against neurons that are sensitive to pressure, sensitive to touch. The fire beetle feels the fire not as heat but as pressure, and it flies toward that pressure, looking for a mate, looking for a place to lay eggs. The fire beetle sometimes successfully finds barbecues and smokestacks and stadium lights. It sometimes lands on firefighters.

The mechanism used by the fire beetle is so sensitive that it has attracted the attention of the military. Duplicated electronically, the mechanism may offer a new approach to infrared scopes of the sort used to see at night, to see through smoke-filled rooms, to find hot spots when mopping up burned-out fires.

Before the fire, deer, coyote, black bear, and an occasional mountain lion roamed the La Brea hills. Mice, kangaroo rats, and shrews would have been here in the brush, along with chipmunks, raccoons, and skunks. There would have been bobcats. There would have been fence lizards, king snakes, and rattlesnakes. The larger mammals, to the extent they could, would have fled before the fire. Some of the smaller mammals would have been baked in trees. Others, along with the snakes and lizards, would have hunkered down in burrows or rock piles, surviving or roasting as a function of wind direction and fuel abundance and moisture and luck.

Birds, once out of the nest, have the advantage of flight. Firefighters sometimes see birds fleeing fires and, on occasion, claim to see them ignite in midair.

Unlike plants, unlike fire beetles, mammals and birds and reptiles do not have special adaptations for surviving chaparral fires. They do the best they can. Because chaparral burns irregularly, skipping patches here and there, survivors find a place to live while the land recovers. Or not. Some of them, having survived the flames, will starve to death for lack of forage.

A government report on chaparral wildlife offers the following words of comfort: "From an evolutionary point of view, however, these deaths are inconsequential."

Burned chaparral, recovering, supports different species than mature chaparral. In the first few years after a fire, cactus mice and harvest mice might come and go. Brush mice may become abundant only in later years. Young chaparral supports more deer mice and coyotes, but fewer dusky-footed wood rats and California mice. Parts of nature die while others thrive.

Predators are known to patrol fire lines, looking for game flushed out by flames or, after the fire, game that is looking for green forage.

※

In 1950, New Mexico's Capitan Gap Fire took seventeen thousand acres. A fire crew, threatened by flames and heat, dug into the soft earth of a recent landslide and covered themselves with dirt. They later emerged alive. Also alive after the fire was a young black bear, a cub clinging to a tree, its hair singed, its skin burned in places. The bear's rescuers named him Hotfoot Teddy. He was later renamed after the advertising bear, the cartoon Smokey. Hotfoot Teddy became the living version of Smokey Bear and as such lived in the National Zoo in Washington, D.C. He lived until 1976, and when he died he was buried at Capitan, under a stone marker.

Smokey's image was painted for a twenty-cent stamp. Restaurants and streets and a historical state park have been named after him. In a single week, Smokey Bear has received as many as thirteen thousand letters. Although he could not read his letters, he was given his own zip code: 20252.

All this notwithstanding, Edward Abbey, critical of Smokey's campaign to control nature, is not alone as a detractor of Smokey. Smokey has been blamed for government policies of fire prevention, of wrongheaded efforts to stop the natural occurrence of fire. Abbey called the bear an idiot, but others have branded him a pariah and worse. Smokey has been blamed for accumulation of fuels in American forests. He has been blamed for the nearly wholesale burning of Yellowstone National Park in 1988, when years without fire resulted in the accumulation of enough fuel to lead to what some saw as an unnatural fire—unnaturally large and unnaturally hot, thanks to the bear.

The former firefighter and renowned fire historian Stephen J. Pyne described the Yellowstone fire in an article in *Natural History*: "Groves of old-growth lodge pole pine and aging spruce and fir exploded into flame like toothpicks before a blowtorch. Towering convective clouds rained down a hailstorm of ash, and firebrands even spanned the Grand Canyon of the Yellowstone. Crown fires propagated at rates of up to two miles per hour, velocities unheard of for forest fuels. A smoke pall spread over the region like the prototype of a nuclear winter. Everything burned."

But Smokey has his place. Some fires, sometimes, can be prevented, or at least delayed. Some fuel loads, sometimes, can be managed. The key to fire management today seems to rely on premature ignition, on setting controlled burns to manage fuel loads and to reset the chaparral. Had Smokey been on the job, the Tea Fire that burned the photographer's house might have been prevented. Maybe, with the right funding and the right planning and the right amount of foresight and luck and unbounded optimism, the fuel load that let flames rip down the hill from the Tea House could have been burned piecemeal, a controlled burn done in manageable swaths to renew the chaparral without destroying homes and overrunning fire engines and melting a file cabinet full of irreplaceable photographs.

❋

Back in Santa Barbara, I stop at the house of the photographer's mother, an artist known for her mosaic murals. The county has commissioned her to create a mosaic commemorating fire survivors and first responders. It is to be a community participation mosaic, art as therapy for people who live surrounded by fuel, for people who live with fire.

The mosaic, under construction, is too big for her studio. She has moved it into her gallery, attached to her house. It is art under construction, laid out in rough form. Its substance comes from used fire equipment and fire. Half of a toy metal excavator sits near the bottom of the mural. It was salvaged from a fire and then sliced lengthwise using a plasma cutter. Now it represents the cutting of a firebreak, the sort of work that was going on at the Spanish Ranch Fire when the bulldozer operator was overrun by flames, the sort of work that forms the very basis of fighting wildland fires, the sort of work that is familiar to the residents of Santa Barbara.

A canvas fire hose stretches across the top of the mural toward a fire artifact figurine, a girl. The girl is surrounded by flames made from dichroic glass, glass that has been heated in the presence of metals, creating a metallic vapor that cools to form a crystal structure on the glass surface. Dichroic glass shimmers in surprising ways, playing with light. The fire artifact girl will be forever surrounded by dichroic flames, forever on the verge of rescue by firefighters.

In another part of the mural sits a glass bottle, deformed by heat, along with a wineglass, dented and twisted, perhaps from the photographer's house, the glass that had been covered by ashes and was still hot four days after the fire.

Edward Abbey, reading Burton's *The Anatomy of Melancholy* in his fire tower, looking out over a landscape that was not burning, would have come across a passage about urban fires. He may have underlined it because of his interest in fires, or he may have ignored it because of his place in the wilderness, far removed from cities. But here, in Santa Barbara, it is relevant. "How doth the fire rage," wrote Burton, "that merciless element, consuming in an instant whole cities! What town of any antiquity or note hath not been once, again and again, by the fury of this merciless element, defaced, ruinated, and left desolate?"

Chapter 3

COOKED

Iam in Rio de Janeiro. Last week, mudslides closed streets. People died, their homes swept away in rivers of mud. But the rain has stopped. The pavement steams. The eroded remains of ancient magma—spires of straight-up rock—tower above buildings and bays and beaches.

In 1992, the United Nations met here. The meeting became known as the Rio Earth Summit. More than a hundred national leaders attended along with nineteen thousand others who tagged along to offer advice. In a two-page preamble, the delegates acknowledged that climate change was a common concern of humankind and that human activities enhanced the greenhouse effect. Most of the carbon emissions, they said, came from developed countries. They divided the world into haves and have-nots, the developed and the undeveloped. The haves would try to hold carbon emissions at 1990 levels. The have-nots would emit at will until they became haves.

Brazil was a have-not.

The twenty-four-page document that came from the Rio Earth Summit contained no binding requirements to reduce carbon emissions. The haves agreed only to provide detailed information on exactly what they would do "with the aim of returning individually or jointly to their 1990 levels these anthropogenic emissions of carbon dioxide and other greenhouse gases."

All of that is ancient history, part of the long confusion of climate change policy. Now I am here on business, attending an oil and gas conference for environmental professionals. It is a week of afternoon cookies and evening beers and endless presentations. The conference is hosted by the Brazilian national oil company, Petrobras. Petrobras is among the world's largest companies, larger by some measures than Shell or Chevron or BP. And they have just found new oil, more than five billion barrels of it, expanding their reserves by 50 percent. It is oil that they would like to bring to the surface, to burn, to convert to carbon dioxide.

A high-ranking Brazilian official talks about climate change, about carbon emissions, and about the five billion barrels of new wealth. The world must cut carbon emissions, he tells the audience of five hundred, but the cuts must come in the developed world. Countries like Brazil need time to develop. They must have the freedom to emit at will.

John Tyndall, Faraday's friend and biographer, working in the middle of the nineteenth century, could not have foreseen the level of concern that greenhouse gases would create. He could not have known that his work on greenhouse gases would interest a movement calling for sweeping changes in energy use. He could not have foreseen the Rio Earth Summit. Tyndall was merely curious about the manner in which air absorbed heat. He wanted to follow up on what was even then known as the greenhouse effect, on Fourier's work showing that the earth was much warmer than it should be, suggesting that the atmosphere was a blanket, a comforter enshrouding the earth. Tyndall wanted to understand how it worked.

He published his findings related to the greenhouse effect in 1861. He started with a review of past work. "So far as my knowledge extends," he wrote, "the literature of the subject may be stated in a few words." In fact, his review of the literature required ninety-five words.

Tyndall's idea was simple enough: compare the heat passing through a tube full of air with that of a tube filled with nothing, a vacuum. "The first experiments," Tyndall wrote, "were made with a tube of tin polished inside, 4 feet long and 2.4 inches in diameter." His heat source was a cubical bucket filled with hot water.

Tyndall could not detect heat absorption in air. "Oxygen, hydrogen, and nitrogen," he wrote, "subjected to the same test, gave the same negative result."

He tried hot copper and hot oil as heat sources. He tried a lamp. "During the seven weeks just referred to," he wrote, "I experimented from eight to ten hours daily; but these experiments, though more accurate, most unhappily shared the fate of the former ones."

He went back to the cubical bucket filled with boiling water. He tried what he called olefiant gas, known today as ethylene, a colorless gas as invisible as air itself. But it was not invisible to radiant heat. The needle on his instruments moved seventy times farther than it had moved for oxygen, hydrogen, and nitrogen. He could not believe his results. He repeated the experiment several hundred times. "I was indeed slow to believe it possible," he wrote, "that a body so constituted, and so transparent to light as olefiant gas, could be so densely opake to any kind of calorific rays."

He tried other gases: bisulphid of carbon, iodide of methyl, chloroform, amylene, chlorine. In the end, he realized the importance of two gases. He wrote that carbon dioxide and water vapor "would produce great effects on the terrestrial rays and produce corresponding changes of climate."

In this remark, he was not thinking of industrial pollution. He was not concerned about the burning of fossil fuels. He was concerned with what he called "the mutations of climate which the researches of geologists reveal." He was concerned with the end of the Ice Age.

Before deforestation could become a problem, before climate change could become something more than a matter of intellectual curiosity, humans needed to control fire. It was a time before charcoal production, before coal mining, before oil and gas drilling, before whale oil lamps, before Faraday's candle and Tyndall's tube, before a gasoline-powered mower sparked against a rock and lit the Spanish Ranch fire that killed Scott Cox and Ed Marty and their two colleagues.

Humans needed the ability to strike a flame whenever and wherever they desired. Exactly when this happened is not known. When this happened is not even known in the approximate sense.

At Swartkrans Cave in South Africa, archaeologists found bones burned in a manner consistent with the use of a campfire. That campfire may have burned just over a million years ago. The owners of those bones may have been *Homo erectus*, a not-so-distant ancestor, an ancient human who, in the right outfit, would not look ancient at all.

Archaeologists have weaker evidence of controlled fires in

Kenya, at a place called Koobi Fora, dated to just over one and one-half million years ago. There are other scattered sites and scattered evidence, bits of what seem to be fired clay, depressions with charcoal, with burned stones. The evidence increases at younger sites. From seven hundred thousand years ago, at the Bnot Ya'akov Bridge in Israel, evidence suggests butchering and cooking. By a few hundred thousand years ago, someone was laying out what appear to be fire rings, what archaeologists call hearths, burned stones in circles of the sort commonly made by Boy Scouts. By a hundred thousand years ago, both *Homo sapiens* and *Homo neanderthalensis* used fire. They used it for cooking and for warmth and perhaps for hunting and warfare.

But they did not have matches. Phosphorus matches, similar to what we know as matches today, appeared around 1832, and modern disposable lighters showed up in force in the 1970s with the Cricket, marketed by a friendly singing cartoon cricket, and the Bic, good for three thousand lights, three thousand flicks of the Bic.

※

I am back home, in a more comfortable climate than the deserts and burned-out chaparral of California, more pleasant than the humidity of Brazil. I am on skis, headed above the tree line in the Chugach Range above Anchorage. In my pack, I carry a fire-starting drill and bow, a block of magnesium and a striking plate, a disposable lighter, and fifteen storm-proof matches. For tinder, I have dried birch bark and fire lighter blocks. For fuel, I have dried and split sticks of birch. The plan is simple: start a fire as ancient man might have started a fire. The matches and the disposable lighter and the fire lighter blocks are backups, in case I prove less adept than *Homo erectus* and dumber than *Homo neanderthalensis*.

I turn left and head up a treeless valley through deep snow, fresh and brilliantly white, an almost perfect reflector of sunlight and heat. The valley takes me gently higher until I am slowed by swollen hills of glacial moraine too steep to conveniently climb on skis. Near the top of one of these hills, wind has exposed bare rock, a rounded and windswept boulder. Next to the rock, I drop my pack and step out of my skis. Without skis, I sink thigh deep in snow.

From my pack, I take my fire-starting drill and bow. The bow is a branch of birch, still green, strung with a thick shoelace. The drill is a very dry and straight poplar branch, a foot long, sharpened at both ends. I have a base of dried pine, notched to hold the drill tip and to catch a hot ember. In the deep snow, I struggle to pin the base to the rock, ultimately leaning down on it with one knee, my body awkwardly stretched. I insert the tip of my wooden drill into the dried pine base, pushing the drill's tip into the base's notch. I wrap the bow's shoelace around the poplar drill so that a sawing motion on the bow will spin the drill. With my left hand, I cup another piece of pine, pressing it down against the top of the drill, and with my right hand I saw, moving the drill back and forth. I move slowly at first, and then faster, seeking heat from friction.

"Savages," wrote John Tyndall in *Heat: A Mode of Motion*, "have the art of producing fire by the skilful friction of well-chosen pieces of wood." And he wrote of Aristotle: "Aristotle refers to the heating of arrows by the friction of the air." And Tyndall wrote of Benjamin Thompson, also known as Count Rumford, who had noticed the heat generated by the drilling of cannon bores. In 1798 Rumford immersed an iron cylinder in water. He used a horse to turn the cylinder against a rod, generating friction, which generated heat. Tyndall quoted Rumford: "It would be difficult to describe the surprise and astonishment expressed by the contenances [*sic*] of the by-standars on seeing so large a quantity of water heated, and actually made to boil, without any fire."

Rumford—a friend of Lavoisier, and the man who married Lavoisier's widow—worked at a time when heat was still believed to be a mystical fluid, the fluid that Lavoisier called caloric. In the beginning, Rumford believed that the heating of one item by another came from the flow of this fluid. But Rumford's experiments showed the cannon producing more heat than it could have held, producing more heat as more motion was applied. "I think," Rumford wrote, "I shall live to drive caloric off the stage as the late Lavoisier drove away the previous theory. What a singular destiny for Madame Lavoisier!" Rumford laid the groundwork that would allow Tyndall to see heat as a mode of motion.

I am not a savage. My drill slips out of the notched base. Both base and drill land in snow. I try again. After five minutes, I see no smoke. I touch the tip of my drill and feel no heat. I touch the notch in my base and feel cold wood. There are two possibilities: Rumford and Tyndall were wrong, or I am incompetent.

From my pack I take my block of magnesium and the striking plate. I shave magnesium onto the wooden base. The idea is to strike a spark into the magnesium, which will flare up suddenly with a very high temperature. With my knife, I strike sparks down into the magnesium shavings. They do not light. I strike again and again until, in one clumsy motion, I knock the base and the magnesium shavings into the snow.

From my pack I take my disposable lighter. I click it again and again. It will not light. It is time to dispose of my disposable lighter.

From my pack I take my storm-proof matches. I strike one and touch it to my birch bark tinder. The match burns out before the bark ignites. I try again. And again. I burn through seven matches before the birch bark lights. And from there, I build a small fire of birch sticks. On top of the fire I place my drill and bow, and then the notched pine base and the notched pine handle. I step into my skis and dispose of my lighter in the flames. A few seconds later, the fire burns through the lighter's plastic casing, and with a disappointing soft pop, the flames consume what little is left of the lighter's butane.

I watch as my little fire pumps carbon into a warming atmosphere, and when the fire dies, I pick up the remains of my lighter, spread the fire's ashes in the snow, and ski away, down the valley, back toward the tree line, happy that no one was watching as I proved myself less artful than a savage, less adept than *Homo erectus,* and dumber than *Homo neanderthalensis.*

If the world were populated only by people like me, we would still be living in trees and eating raw fruit. Climate change would not be an issue.

❉

It is possible to buy, for under one hundred dollars, a mint condition book of matches from World War II with a cover that says, "Strike 'Em Dead; Remember Pearl Harbor." Each match is printed with the figure of a Japanese soldier. When lit, the tiny soldiers burn from the head down. At today's prices, it would cost several dollars to start a fire with a vintage Strike 'Em Dead match.

The modern match had many predecessors. One predecessor, called the "promethean," was patented in 1828. It was a glass bulb filled with sulfuric acid. The bulb was wrapped in paper coated with potassium chlorate. Users could crush the bulb between their teeth to ignite the match.

At about the same time, the combination of sulfur and phosphorus was commercialized by an English pharmacist named John Walkers. He sold yard-long "sulphuretted peroxide strikables." Aside from their inconvenient size, strikables had an undesirable tendency to self-ignite. By 1832, smaller versions were available, but the self-ignition problem remained. There were issues, too, with the manufacturing process. Women working in factories suffered from "phosphorus disease" or "phossy jaw." The handling of white phosphorous turned the sides of their faces green. Their jaws wasted away and exuded pus. Their bones reportedly glowed with an eerie green light. Tissue necrosis brought fever and eventual death.

Twenty years later, red phosphorus replaced white phosphorus, and the invention of the safety match ended the problem of self-ignition. The match head contained half the formula, and a striker plate contained the rest.

In 1889, the pipe-smoking lawyer Joshua Pusey made paper matches and called them "flexibles." A few years later, the Mendelson Opera Company advertised on the covers of books of flexibles: "A cyclone of fun—powerful cast—pretty girls—handsome wardrobe—get seats early." Cast members may have printed the advertisements by hand. This led to the Strike 'Em Dead match and, in the European theater of war, a book of matches printed with instructions for derailing German trains. Matchbooks advertising restaurants and bars and hotels were inevitable, as was the birth of the American Matchcover Collecting Club and the opening of the Match Museum, self-proclaimed as the world's only match museum, in Sweden. It was merely a matter of time before someone used matchbooks to promote cancer prevention and discourage smoking. The matches inside these matchbooks, like my bow and drill, do not light.

I telephone the Firewalking Institute of Research and Education. I ask the instructor what sort of wood is used on fire walks. "Cedar," he says. "Always cedar." Cedar crackles. It has a pleasant smell. Cedar is the obvious wood of choice for firewalking.

A typical fire walk burns about a quarter cord of cedar. A

neatly stacked cord measures four feet high by four feet wide by eight feet long. A cord of cedar might contain twenty trees, or it might be a single tree, or only part of a single tree. The number of trees per cord depends wholly on the size of the trees. To walk on fire might be to walk across five burning cedar trees, or one, or only part of one.

A cord of dry cedar, burned, releases roughly the same amount of heat as 164 gallons of gasoline, burned. Burning a quarter cord of dry cedar would be comparable to burning forty-one gallons of gasoline. But gasoline burns far faster than wood. Gasoline releases its heat in a few almost explosive moments. To firewalk through forty-one gallons of gasoline would be to exceed the boundaries of prudence.

"It is important," the instructor tells me, "for firewalkers to ignite the fire, or at least to see it lit. They need to understand that this is real fire. These are real flames and real coals."

The cedar fire is struck with a lighter of the sort sometimes used to light a backyard barbecue. Kerosene might be used as an accelerant. "Never gasoline," he says. He is adamant. "Never, ever, gasoline."

Another safety tip: never cook marshmallows over a firewalking fire. Walking across hot coals is one thing, but molten sugar adhered to the feet is another.

Rain falls in the Anchorage suburbs. As I often do, I build a fire in my woodstove. My source of ignition is a novelty lighter in the shape of a double-barreled shotgun. Flame shoots from both barrels. A single wad of the *Anchorage Daily News* burns. The news ignites finely split and dried birch. The kindling ignites sticks of fuelwood, sixteen-inch lengths, mostly birch, with the occasional stick of poplar. One of the few failings of Alaska is the paucity of good firewood. We have no forests of ash, or oak, or hickory. On the other hand, we do not have to suffer through the ineptitude of sequoia or fig.

The fire forces water from the wood. A living tree is perhaps 50 percent water. Seasoned firewood holds about 20 percent water. Water does not burn. The separation of water molecules, the conversion of liquid to vapor, requires energy. It steals the heat from my stove. Wood steams in the young fire, the fire just after ignition, the fire that exists while the newspaper still burns.

The fire takes hold, its heat forcing cellulose molecules to dance. Pure cellulose is odorless, tasteless, and white. Cotton is nine-tenths cellulose. Reasonably dry wood is about half cellulose. Cellulose is six parts carbon, ten parts hydrogen, and five parts oxygen, the various parts forming rings, the rings themselves joined together one after another after another. A single cellulose molecule with a thousand rings joined together in one long chain would be at home in the cell wall of a tree.

The chains are tough. To burn, they need to be heated to the point of dancing. With heat, they fox-trot, they gyrate, they twirl, they cha-cha-cha, they twist. They do the Lindy Hop, the lock, the pop, the toprock, and the downrock. They dance with such vigor that they occasionally shed parts. They break. A carbon snaps off. In the updraft, in the convection storm rushing toward the chimney, two oxygens intercept the carbon, and the three parts join to become carbon dioxide. This is barroom dancing, where changing partners is part of the routine, where dancers may be alone or in pairs or in groups of four or five or a dozen. A hydrogen snaps off, and another, and two more, and the four hydrogens join a carbon to become methane. The methane finds more oxygen and reinvents itself as water and carbon dioxide. In the flames, another methane forms, picks up more partners, and becomes propane and then butane and octane, short-lived but lively. The dance floor rocks, crowded and chaotic.

My woodstove, in burning wood, makes and burns ethane and propane and butane and even pentane and heptane and octane, all of it coming and going, the net result a rising tempo, one dancer spurring on the next, all yielding heat.

The flame's appearance is as it is, in part, because of glowing particles of carbon, of carbon heated to incandescence. The temperature affects the color. But the flame's appearance is as it is in part because of impurities in the wood. Sodium shines yellow. Potassium shines purple. Copper shines green.

The fire pops. Pockets of steam or hydrogen or methane or pentane explode through the wood. Explosions send up sparks.

Ash accumulates in the bottom of my stove. Ash is sodium, copper, sulfur, silicon, and potassium. Good wood burned in a good fire yields little ash. A quarter cord of cedar weighs seven hundred pounds as wood but only four pounds as ash. The lesson: if you have to haul firewood, burn it first.

In my woodstove, I have created a firestorm. If I crack open the door, air rushes in. The fire flares. Flames move sideways and to the back of the firebox and up. But flames move downward too, extending three inches beneath a burning stick of birch before curling and heading upward, toward the chimney. The fire in the woodstove creates its own weather. That weather is not pleasant. In miniature, the wind coming through the cracked woodstove door is a Santa Ana, a sundowner, a wind ripping into an accumulation of fuel.

My stove is of the type that is sometimes called a Franklin stove. Ben Franklin would not recognize it as such. In a self-published pamphlet, he described his stove in 1744. He was motivated by heat rather than profit, by his sense of duty to society, by frugality. The fire, in the standard fireplace of Franklin's time, sucked warm air from the room, heated it, and sent it up the chimney. "The upright heat," Franklin wrote of a standard fireplace, "flies directly up the Chimney. Thus Five Sixths at least of the Heat (and consequently of the Fewel) is wasted, and contributes nothing towards warming the Room."

The warm air from the room was replaced by cold air from outside that leaked in through every door and window and crack and crevice. The cold drafts, Franklin believed, caused illness. "Women particularly," he wrote, "from this Cause, (as they sit much in the House) get Colds in the Head." He could not abide wasted wood and sick women, so he built a better stove. His stove looked in some ways like the modern woodstove, a metal box in which wood burns, but Franklin's stove was open on the front and contained baffles that circulated heat and smoke before sending it up the chimney. Franklin's stove, by modern standards, was clumsy and less efficient than the modern woodstove, but it offered twice the heat of a fireplace and used less wood.

Franklin's detractors claimed that iron stoves smelled. Franklin blamed the smell on spitting. "To spit upon them to try how hot they are," he wrote, "is an inconsiderate, filthy unmannerly Custom; for the slimy Matter of Spittle drying on, burns and fumes when the stove is hot."

Franklin was asked to patent his stove. He refused. "As we enjoy great advantages from the inventions of others," he wrote, "we should be glad of an opportunity to serve others by any invention of ours; and this we should do freely and generously."

I am alone. No one can see what I do. I spit on my stove and watch the spit turn to steam. I smell nothing. Franklin's spitters, I conclude, may have been tobacco chewers.

Into my stove I toss a chemistry textbook. I watch the cover melt. The fire reduces twelve hundred pages to heat and ash and smoke. The formula for cyclohexyl methyl ether burns, as do those of various aldehydes and ketones and esters. The molecular structure of allylic free radicals, illustrated by stick figures, burns. "Heating oil," the book says, "is a mixture of hydrocarbons in the 14 to 18 carbon range." The sentence burns. Entire processes go up in smoke: reactions of epoxides and alkenylbenzenes, conversion of alcohols to ethers, protein synthesis.

People who heat their homes with wood are fond of saying that their fuel warms twice. "They warmed me twice," wrote Thoreau in 1854, "once while I was splitting them, and again when they were on fire." But this is an incomplete statement. Wood warms at least three times. The third time it warms the earth. A quarter cord of cedar becomes something like two thousand pounds of carbon emissions. My four-pound chemistry textbook, burning, becomes something like thirteen pounds of carbon emissions, each of its carbon atoms weighed down by two oxygen atoms that more than triple its weight. But now it is mostly gas, out of sight, its intellectual content reduced to ash, steam, soot, and carbon dioxide.

Zoroaster was a man of the Bronze Age, from a time well before Christ. He believed in a single benevolent god, in a world where good had to fight to prevail over evil, a world in which humans could actively support the side of good through appropriate thoughts and words and deeds.

The Zoroastrians built a myth of three Great Fires that have existed throughout time. They built domed fire temples with actual fires. They maintained the fires, which involved a steady supply of fuelwood and prayer. The fires could be moved from place to place as the exigencies of history required. In the tenth century, the Zoroastrians moved from the Middle East to India, and they may have carried a sacred fire with them, or at least its ashes. Through fire, they believed, one could find wisdom.

In India, they became the Parsis. From there, they spread further afield. Parsis, and therefore Zoroastrians, live in scattered communities throughout the world. The Londoner Farrokh Bulsara, also known as Freddie Mercury, front man for Queen, composer of "Bohemian Rhapsody," "Somebody to Love," and "We Are the Champions," was of Parsi descent. His album *Queen on Fire* was recorded live during the 1982 Hot Space Tour.

Freddie Mercury, when alive, was not the only Zoroastrian in London. There were, in fact, enough Zoroastrians in London to justify a fire temple in West Hampstead.

I am neither a Parsi nor a Zoroastrian, but I am intrigued by the possibility of moving fire from place to place, an eternal flame, a torch, an elimination of the need to rub two sticks together. People once carried embers in pouches of various kinds. The Iceman of the Alps, who lay frozen near the Italian-Austrian border for five thousand years before being found by German hikers, carried embers.

The Iceman's ember pouch was made from birch bark. I make mine from a scrap of deer hide lined with birch bark. From my fire, I scoop an ember, two inches in diameter and three inches long, glowing red. I drop it into my pouch, then cinch the top closed, cutting off the air supply. I set it aside until the following morning. I open it. The ember is black and cold, reduced to a frigid lump of charcoal, as useful for starting a fire as a drill and bow in the hands of an incompetent suburban skier.

I would fail as a Neanderthal, I would fail as a Zoroastrian priest, and I would fail as an Iceman.

The Asháninka people live in parts of Brazil and Peru. An Asháninka cookbook might include recipes for meat cooked in a monkey's stomach and river turtle eggs cooked with turtle meat in the turtle's shell. And there is frog boiled in bamboo: gut freshly caught frogs, wash and salt the carcasses, place the salted carcasses and a bit of water in a pipe of bamboo, seal the bamboo pipe with herbs, and heat to boiling over an open fire. Serve warm with boiled manioc or cassava or yucca, or with roasted bananas.

The question is this: how did the ancestral ape as a raw food-ist become a cooking ape, an ape capable of controlling fire and cooking elaborate dishes, haute cuisine like frog boiled in bamboo and chocolate mousse and bouillabaisse and andouillette and seafood paella with a squid ink sauce?

It may have been the control of fire first, and then cooking. The ancestor may have found comfort in the light and warmth that came from the fires of lightning strikes and lava flows, and he may have nourished that fire when he could, and at some point found it useful for cooking. Or the ancestor may first have found fire because of its ability to cook prey.

The ancestor's teeth—his dentition—show that he was omnivorous, an eater of leaves and roots and fruits, but also of raw bugs and raw mammals and raw fish. The ancestor was hungry. And he was resourceful. He could see a forest burning, flushing before it injured animals, half cooked, and he could find burned-over carcasses, fully cooked.

Modern apes and monkeys will approach fires. In captivity, they handle burning sticks. They slam burning brands against the ground to send up sparks. Some smoke cigarettes. Whatever innate fear of fire exists can be overcome with the right motivation. They will also eat, preferentially, cooked food. They like cooked carrots and potatoes and meat. Even chimpanzees that had never before eaten meat were found to prefer it cooked to raw. What is true of modern apes was likely true of ancient man.

Homo habilis—named by Louis Leakey to mean "handy man"—wandered the earth just over two million years ago, among the first of the genus *Homo*, among the first that science recognizes as being one of us or nearly one of us. His teeth were smaller and his brain was bigger than those of his more primitive ancestors, the *Australopithecines*. He may have spent a substantial amount of time in trees, but he also made tools. *Homo erectus* came next, Peking man, walking upright with locking knees less than two million years ago. And then came *Homo sapiens,* a few hundred thousand years ago, his skull expanded, his ability to make tools and grow crops and build cities and manufacture bombs uncontested. But all of this oversimplifies the truth. The truth carries the burden of extinct cousins, lines that were human, more or less, but died out, or lines that might have been transitional between other lines. There is *Homo antecessor,* found in Spain and England, a bit more than a million years old, and *Homo cepranensis*, known from a single skullcap found in Italy, about five hundred thousand years old, both maybe somehow linked to *Homo heidelbergensis,* from around six hundred thousand years ago, who may have been the final link between *Homo erectus* and *Homo sapiens*. *Homo heidelbergensis* may also have been the link to *Homo neanderthalensis,* Neanderthal man, an offshoot that went extinct, a cousin rather than a grandfather, but a cousin very handy with tools and fire, a cousin who interbred with early *Homo sapiens*, with Cro-Magnon man.

By the time *Homo sapiens* showed up, fire was widely used. *Homo erectus* had used fire in Swartkrans Cave and Koobi Fora. The hand axes and bones found amid burned olives and barley and grapes at Bnot Ya'akov Bridge come from the time of *Homo erectus*. By the time of *Homo sapiens,* fire was part of everyday life. *Homo sapiens* was born halfway down the on-ramp to climate change.

※

Richard Wrangham, a Harvard primatologist, an ape studying apes, argues that fire and cooking drove human evolution. The emergence of cooking led to the emergence of modern man, the rapid changes that characterized the road from *Homo habilis* to *Homo sapiens*. Cooking provided access to otherwise inedible stuff. Tough meats became chewable, hard shells cracked, and certain poisonous plant compounds were neutralized. Maybe more importantly, cooking increased the energy yield of food. "Cooking," Wrangham wrote, "increases the amount of energy our bodies obtain from food." Drippings lost to the fire were made up for in the digestive process.

Take meat as an example. Connective tissue makes it tough. Connective tissue is collagen and elastin, fibrous protein and rubbery protein. It is skin and fat and the stuff that holds the body together in neat little compartments. Heat collagen, and it falls apart. At around 150 degrees, collagen molecules lose their shape, they unravel, they melt, they become Jell-O. They become easier to cut, easier to chew, and easier to digest.

The body extracts most of its useful energy supply from food in the stomach and small intestine. Undigested food finding its way into the large intestine intact is of little value. An ape eating cooked food has an advantage over an ape eating raw food. The cooking ape produces more offspring than the raw-foodist ape.

Charles Darwin once described cooking as the second most important discovery after that of language.

From Jean Anthelme Brillat-Savarin, author of *The Physiology of Taste*, a book published in 1825 that remains in print today: "A man does not live on what he eats, an old proverb says, but on what he digests."

From Richard Wrangham: "The introduction of cooking may well have been the decisive factor in leading man from a primarily animal existence into one that was more fully human."

❋

A quarter million years ago, someone left thick ash and hearth stones in a cave in Israel. That someone may have been *Homo erectus* or *Homo heidelbergensis*. A hundred thousand years later, someone cooked an elephant, leaving behind charred bones and ash. This someone was *Homo sapiens,* or his cousin *Homo neanderthalensis,* species that evolved from the beginning with a knowledge of fire and cooking.

Seventy-six thousand years ago, someone near Barcelona left behind sixty hearths, burned bones, and imprints of long-gone wooden cooking utensils. Forty thousand years ago, someone in France's Dordogne region dropped fire-heated stones into water, a method of boiling water in the absence of fire-resistant pots.

We called ourselves *Homo sapiens,* "the wise man," but we may more aptly have called ourselves *Homo crustulum,* "the cooking man." Our brains grew larger, but our teeth shrank, our intestines grew shorter, and our jaw muscles grew pathetically weak.

After all of this, we find ourselves with the science of molecular gastronomy, the cross between the laboratory and the kitchen. It took a physicist and a chemist, one Hungarian and the other French, to turn food science into molecular gastronomy. They did so through seminars and presentations. They held workshops on texture, on chewing, on swallowing, on foams made from egg whites that go by the name of mousse, on dough.

The physicist was Nicholas Kurti, and the chemist was Hérve This. Starting in the 1990s, they brought scientists together with chefs. They put graduate students in kitchens. They asked questions. What happens when you roast meat? Does slow heating make the best broth? What is the perfectly cooked egg? They held seminars on the methods of heating: by liquids, by steam, by air, by radiation. They quoted Count Rumford, the man who married Lavoisier's widow, who showed that friction generates heat, that the heat released by the boring of a cannon could boil water: "In what art or science could improvements be made that would more powerfully contribute to increase the comforts and enjoyments of mankind?"

The science of molecular gastronomy asks how something works, and it asks why something works. Cooking is the work itself. "We shouldn't confuse science with technology," Hérve This once said. "The application of that knowledge is the cooking part," he said, "and that's technology. Cooking is a technique, combined with art."

In Hérve This's world, the boiling of an egg becomes a matter of scientific interest and technical importance. As the egg warms, proteins unravel and then coagulate into a molecular net, a mesh, but the proteins of egg whites and egg yolks unravel and coagulate at different temperatures. Use different temperatures and cook different eggs, with different consistencies. The time of cooking is not important. As long as the temperature is controlled, the three-minute egg is no different from the five-minute egg or the twenty-minute egg. Hérve This talks of eggs of different degrees. For example, he might talk of a 149-degree egg with a soft orange yolk, or he might talk of a 153-degree egg with a firmer yolk, yellow rather than orange.

A journalist once recorded This spotting an undercooked egg in a restaurant. It was not that the egg was inedible but that it was a 147-degree egg sold as a 149-degree egg.

I throw a slab of halibut, freshly caught and filleted, onto my grill. Halibut, undercooked, has an unacceptably mushy texture and the taste of raw fish. Overcooked, it is tough and tastes of cardboard. The difference between undercooked and overcooked is a matter of seconds. Hérve This has not commented on the cooking of halibut, but he would certainly mention the importance of timing in the coagulation of proteins. He would mention, too, that cooking removes water from the flesh. And he might suggest something about Maillard reactions, the same reactions that color bread crust, give beer a golden luster, and darken cooked meat, reactions involving amines and Schiff bases and Amadori products that react with other compounds to become aromatic molecules—molecules in the shape of a ring that sometimes come with an odor. But in all the chemistry, Hérve This would not lose sight of the technique and art of cooking. I flip the halibut twice and remove it just in time, within seconds of disaster, but, in this instance, perfectly prepared and ready to increase the comforts and enjoyment of mankind.

※

In 1957, Thomas E. Briscoe told arresting officers that he had been setting fires since he was twelve years old. By his own estimate, he had set more than a hundred fires. In the past year alone, he had set more than fifteen fires in the area where he was arrested, and the fire that led to his arrest was his second fire that night. According to his confession, he would wake up at night feeling what court documents referred to as a "strong sexual urge." To satisfy the urge, he would light a fire. He would call the fire department and watch them work, and afterward he would obtain "sexual gratification, sometimes with his wife and sometimes by masturbation."

The arresting officers were quoted as saying that Briscoe "was sick and needed psychiatric treatment." One examining psychologist said that Briscoe "was suffering from a severe degree of mental defect." Another found Briscoe to be a "borderline mental defective" but not "strictly a mental defective." The legal question was not one of mental defectiveness but one of mental competence. Could his guilty plea be taken seriously? Did he know what he was doing when he lit the fires? Did he know what he was doing when he pleaded?

The psychologist's report declared him competent. Briscoe was "not suffering from mental defect of a sufficient severity to render him mentally incompetent." He had probably been "of sound mind" when he lit the fire and when he confessed.

A confounding factor: Briscoe's intelligence quotient, his IQ, measured fifteen years earlier, was somewhere between 41 and 46. An IQ as low as Briscoe's has been described by various terms and phrases over the course of time, many of which came into the lexicon with good intentions but later became stigmatized and derogatory: cretin, imbecile, moron, feebleminded, trainable but not educable, and retarded.

In an appeal, Briscoe withdrew his guilty plea. The change in plea led to legal machinations but not a trial. Briscoe sat behind bars. "It is even possible," wrote the judge involved with the case, "that the unjust sentence will have been served before decision of the appeal." Injustice was served, not only on Briscoe but on society. "Since release from the penitentiary is generally based on passage of time rather than fitness for release," wrote the judge, "appellant, if he has the pyromania to which he confessed but which was not shown because he was denied a trial, would again imperil the community."

Pyromania combines two Greek words: *pyro* for "fire" and *mania* for "loss of reason." The pyromaniac is crazy for fire. He or she—but usually he—has problems controlling impulses. In a moment of inspiration, the mind says, "let's light a fire," and the pyromaniac, lacking impulse control, follows through. In this sense, pyromania may be related to any number of destructive behaviors. Among these are kleptomania, pathological gambling, mood disorders, unexpected verbal and physical assault, and nail biting. Another is trichotillomania, the pulling of one's own hair.

Sigmund Freud associated pyromania with childhood bedwetting. Whether or not the young Thomas E. Briscoe wet his bed is not known.

Setting fires is more common among the young—kids who

are six or seven or eight years old. Some of them do not necessarily understand the potential danger. They do not feel a sense of sexual excitement. They are not pyromaniacs so much as curious kids with matches and lighters and sometimes cans of gasoline.

Another Briscoe, a woman named Rosa Briscoe, probably unrelated to Thomas E. Briscoe, opened a store in Mississippi in 1981. Things did not go well. In 1982 she allegedly told an acquaintance that "the store needs to burn and it needs help." The acquaintance allegedly found an arsonist. Rosa allegedly agreed to a $5,000 fee. The arsonist taped kitchen matches and sandpaper to the telephone bell in the back of the store. He loosened a gas pipe. A phone call ignited the gas.

Rosa Briscoe may have been guilty of arson and stupidity, but she was not a pyromaniac. She felt neither emotional nor physical arousal. She was fascinated neither by the fire itself nor by the burned building. She did not light one fire after another.

In a momentary lapse of impulse control, I stab the base of a match into a cork so that the cork forms a stand. I light the match. Match and cork go into my microwave oven. I close the door and push the button. Eight seconds later, a flare flashes across the inside of my oven, a mix of fire and lightning, really a plasma, carbon stripped of electrons, with the buzzing noise of sudden ionization, of an electric chair. It is a phase change, akin to ice becoming liquid water and liquid water becoming steam, but here the phase change is from gas to plasma. It is a phase change that makes me smile and laugh, but I have no sense of emotional or physical arousal, no sexual urge at all.

Afterward I question my own intelligence, reminding myself that I am an adult and should behave accordingly, and hoping that I have not ruined my microwave oven.

※

Microwave ovens make certain molecules spin. To spin, the molecules must have an uneven distribution of charge. Their electrons must be crowded toward one end or the other. Water has just such an uneven distribution of charge. Its electrons crowd around its oxygen atom. Its two hydrogen atoms poke out like positively charged pigtails.

Imagine lines in a powerful magnetic field. Place a bar magnet in that field, and it lines up parallel to the field. Now reverse the direction of the magnetic field. The bar magnet turns around, realigning itself. In a microwave oven, the microwave radiation is the magnetic field, its direction reversing two and a half billion times each second. The water molecules are the magnets. The magnetic waves are perhaps five inches long, peak to peak, and they travel at the speed of light.

In an active microwave oven, water molecules spin this way and that. They pirouette. They reverse pirouette. The dancing is heat. It is the same heat felt from the sun or from a fire or from friction.

The microwave oven came from radar. Specifically, the microwave oven came from the radar's magnetron, the component that generated the waves used to detect enemy aircraft.

Soon after World War II, a man named Percy Spencer was working in a Raytheon factory that built radar systems. He carried a chocolate bar in his shirt pocket. A magnetron melted his chocolate.

Spencer returned with popcorn. He popped popcorn in front of the magnetron. He returned with an egg. He blew up an egg in front of the magnetron.

Spencer talked to Raytheon's patent lawyer. A hundred thousand dollars later, Raytheon had a prototype. They marketed it as the Radar Range, selling the first unit in 1947 for three thousand dollars. It was too big for home kitchens, so they sold it to restaurants and hotels. They redesigned it. They got it down to 750 pounds and then passed the torch to Hotpoint, Westinghouse, Kelvinator, Whirlpool, and Tappan. In 1955, Tappan released a model the size of a conventional oven. By 1966, ten thousand Americans owned microwave ovens. Foods packaged for microwave ovens appeared. The ovens grew smaller. By 1970, when Spencer died, forty thousand microwave ovens were sold each year. By the beginning of the twenty-first century, microwave ovens were in virtually every home in America.

The molecular gastronomist Hérve This has written of cooking in Spencer's invention: "Cooked in a microwave, beef is rejected by taste testers, who find fault with its grayish external color, its toughness, its lack of succulence, and its bland taste." The microwaves, according to This, penetrate the surface of the meat, heat the water molecules within the meat, and create steam. The temperature never goes above the boiling point and is never hot enough to trigger Maillard reactions, the reactions of browning. "The oxymyoglobin," This wrote, "is not denatured and retains its color."

But he does not oppose the use of microwaves for certain foods. They can be used, he advises, for poaching fish. And he says they are good for cooking eggs: "Placed in a bowl without an ounce of fat, an egg will cook rapidly; its taste is acceptable, and the figure benefits."

I forgo the use of a bowl and ignore This's implied advice regarding the need to remove the egg from the shell before cooking with Spencer's machine. I place the egg in the middle of my microwave oven. I decide on a three-minute egg. But the egg has other ideas. At forty-three seconds, it explodes. It is a minor explosion, a disappointing pop. The door of my oven remains intact. The eggshell is split, exposing a reasonably well-cooked yolk surrounded by egg white of varying consistency. I do not have the eye of Hérve This, but to me this looks like a 158-degree egg, with hints of a 153-degree egg around the outer edges of the white, most of which are spattered on the walls of my oven.

✳

Faraday, between ignitions, echoed Lavoisier, explaining that food is a form of fuel, that "we may thus look upon the food as the fuel," and that humans, in eating, produce "precisely the same result in kind as we have seen in the case of the candle." It was a realization that he considered "beautiful and striking."

"We consume food," he wrote. "The food goes through that strange set of vessels and organs within us, and is brought into various parts of the system, into the digestive parts especially; and alternately the portion which is so changed is carried through our lungs by one set of vessels, while the air that we inhale and exhale is drawn into and thrown out of the lungs by another set of vessels, so that the air and the food come close together, separated only by an exceedingly thin surface; the air can thus act upon the blood by this process, producing precisely the same results in kind as we have seen in the case of the candle."

He extended the explanation to other species: "All the warm-blooded animals get their warmth in this way."

The heat of the warm-blooded mammal is the heat of the sun. Grass intercepts a bit of energy from the sun. It uses that energy to combine carbon dioxide and water to form a molecule of sugar, a carbohydrate. That molecule holds the essence of sunlight, ready to discharge it later. That sugar, that carbohydrate, has just become a stick of metabolic firewood.

A hen eats the grain from that grass, imbibing its stockpile of metabolic firewood. The hen lays an egg. Hérve This boils the egg with precision and eats it. From grass to This is a chain of who ate whom, a chain of burned firewood, a chain of reactions that are at their most basic no different from those of a burning candle or a camp fire. Carbon combines with oxygen to release energy and carbon dioxide.

When eating eggs, the distance between Hérve This's mouth and the sunbeam is a distance of two links. But in these two links, the vast majority of the sun's energy disappears, wasted to the inefficiencies of life, undigested or lost as waste heat in the chemical reactions of metabolism. Each link represents an 80 or 90 percent loss. The hen recovers only 10 percent of the energy in the grass seed, and This recovers only 10 percent of the energy in the egg, receiving only a single percentage point of the energy originally donated by the sun.

But said another way, Hérve This, in consuming an egg, dines on the sun.

※

John Tyndall, at the end of his 532-page *Heat: A Mode of Motion*, included a few hundred words written by Frederick Boyle. "Among some of the Dyak tribes," wrote Boyle, "there is a manner of striking fire most extraordinary." Boyle describes the Dyaks using bamboo to strike a fire, not by friction but by compression. A bamboo tube is somehow thrust down over a block containing tinder. The sudden heat of compression lights the tinder, like the downward stroke of the piston in a diesel engine igniting its fuel. I find the description difficult to follow and unbelievable. "I must observe that we never saw this singular method in use," Boyle wrote, "though the officers of the Rajah seemed acquainted with it."

Through the mail, I order a fire piston. It is a two-part device, six inches long, distinctly phallic in shape, made from the horn of a water buffalo. One part is the cylinder, and the other part is the piston itself. The piston slides into the cylinder. At the tip of the cylinder is a chamber. The chamber holds a sliver of tinder not much bigger than the head of a match. The instructions: "Push or strike the piston with your palm, driving it forcefully into the cylinder."

The fire piston goes into my pack, along with two kinds of commercially available tinder made from fibers impregnated with wax or oil, and a brand-new disposable lighter. I also pack a small sheet of thin cotton that has been heated to drive away its moisture and most of the oxygen and hydrogen of its cellulose, making it into a cloth of almost pure carbon, a charcoal cloth, a charcloth. Charcloth is the tinder of choice for use in the fire piston.

The plan: start a fire without matches to vindicate myself, show *Homo erectus* and *Homo neanderthalensis* what fire starting is all about, use a Dyak trick to create a flame. I drive an hour north of Anchorage and head up a trail that runs along Eagle River, toward its glacial source. But for the remains of thick drifts, the snow here is gone. The ice that covered the river two weeks ago has reverted to liquid and flowed away, to be replaced by more water from the glacier upstream, flowing in light whitewater around gravel bars and beside gravel beaches. Beyond the beaches, new buds of willow, poplar, and birch turn the landscape green. Higher up, Dall sheep graze, spread across almost vertical terrain. Sunlight warms my face and sparkles on the water, but a cold wind blows downriver, coming from the glacier in gusts.

I find reindeer moss along the trail, a species of lichen in the *Cladonia* genus, good for fire starting. I find dead spruce twigs covered with another lichen, this one sometimes called old man's beard, in the Parmeliaceae family. I find twigs of dead and dried birch and branches of dead and dried poplar.

Three miles upriver, I step onto a gravel bar and drop behind a natural berm, a gravel and cobble wall three feet tall, carved by ice and floodwater. Beneath the berm, I build a small hearth of river rocks, flat rocks forming a base surrounded by round rocks forming a fire circle. On this river bar, there is no fuel other than what I carry. I pin my reindeer moss tinder to the hearth, using another rock to hold it in place, ready for a hot ember from my fire piston. I wad a piece of charcloth into the piston's tinder chamber. I strike the piston with my palm, driving it forcefully into the cylinder. Nothing happens. I do it again with the same result. And again. After five failed strikes, I change out the charcloth, leaving threads of it protruding past the end of the tinder chamber, inviting flame. Five more strikes, but no ember, no glowing tinder.

I look closely at the piston, and the wind grabs the charcloth away, so I load another piece. My striking hand begins to ache. But strike fifteen yields a glow. I have ignition, charcloth glowing from the heat of compression.

I blow on the charcloth, then use the tip of my pocket knife to transport it onto the pile of reindeer moss. The ember smokes for a moment and then falls through the reindeer moss and into a crack between the rocks. My success fails.

I try again with new charcloth. If there is a trick to this, it is a trick that I do not understand. Or it is the trick of mindless repetition. I lose count after ten more failed compressions and two more bits of charcloth.

Then I have ignition again, a glowing ember. I drop the ember onto the reindeer moss. It makes the most smoke. The glow brightens, and the moss smoke thickens. But then the glow fades, and the smoke disappears. Another successful failure.

Giving up, I reach into my pack for my disposable lighter. This one is brand-new and full of butane. Boyle would be ashamed. Tyndall would be amused. *Homo erectus* and *Homo neanderthalensis* would question evolutionary theory. I change my mind. I put the lighter back in my pack.

I strike the piston again and again. A bruise forms on my palm, but I keep striking, driving the piston forcefully. Another glowing ember emerges. I kneel in front of my hearth with my nose inches from the pile of reindeer moss. Carefully, like a pharmacologist, I scoop the charcloth ember into the moss tinder. I gently blow. Smoke rises and thickens. And then, in a sudden flash, the moss ignites. I pile on old man's beard and twigs of black spruce. The old man's beard flares, and the twigs burn. I have a tiny fire. I feed in more twigs of birch and add branches of poplar.

I sit by the fire for a time, satisfied, knowing that the only difference between me and a Dyak tribesman is the need for mail order.

I generate something like twenty-two pounds of greenhouse gas emissions.

I eat peanut butter crackers, uncooked, while the smoke saturates my clothes. I reach into my pack and find my disposable lighter. Although it is brand-new, I dispose of it now, in my small campfire. I step back ten feet, then another ten feet. I watch the fire. I hear a sizzling noise and then a loud pop. A fireball explodes from the hearth. The remains of the lighter itself fly past my head.

A small disposable lighter of the kind that I just destroyed contains a tenth as much energy as a small stick of firewood, but it goes up in a flash, a single bang, its heat emitted as a ball of flame.

I return to my fire. The lighter's warning label, blown off when the lighter ignited, is stuck to one of the rocks forming the fire ring. The label is charred but readable. "Never puncture or put in fire," it says, and "Keep away from children."

Chapter 4

MY CHILDREN EAT COAL

An hour north of Eagle River, weeds grow over steep mine cuts, sparse green fur on top of paradise lost. Erosion gullies are dry now, in July, but two months ago they would have run over with meltwater as the sun ate the snow. Moose tracks crisscross the slopes, along with piles of moose scat. Grizzly tracks cross the dirt trail on which I walk. A golden eagle soars overhead.

Off the trail, the mine cuts rise steeply. To walk upright is to fall. Progress upslope requires a crawling posture and the use of hands.

With my body stooped over and the sun at my back, my shadow falls across fossil plants. It is not a matter of a fossil plant here and a fossil plant there. Fossil plants litter the ground, imprints of unburned fuel. Here I find a piece of gray mudstone, flat, the size of a dinner plate, embossed with the outline of what looks like a poplar leaf and another of a sequoia.

Here is more mudstone, this piece embossed with the stem of an equisetum. Here is a sliver of petrified wood and a branch of the same.

This is what is left of an Eocene forest, a fifty-million-year-old relic. *Eocene:* literally, the new dawn, a time after the loss of the dinosaurs, a time during which birds and mammals diversified and became obviously prevalent, a time when warm-bloodedness and its partner, fever, came into their own, the time of the Pakicetus, the Ambulocetus, the early whales.

Some of the fossils are crumbly aggregations of leaves. Others are more solid, with outlines of stems and branches. Some are bound to chunks of shining black coal. At the top of the mine cut, sweating from the climb, I find a petrified tree trunk, three feet in diameter, a place to sit.

Eocene plants captured the sun and stored its energy, batteries charged by a solar cell. A river ran through here, and intermittently, pools of water formed. Leaves fell from the plants and accumulated, along with twigs and stems and roots and whole trees, settling under shallow water. They became peat, a compressed mass of plant remains, a black garden mulch that forms when moisture blocks out oxygen. Mud and sand, carried in by flowing water, covered the peat. Over time, the weight of mud and sand squeezed out the water, and the peat became lignite, sometimes called brown coal or rosebud coal, a low-grade fuel stranded between peat and real coal but sometimes used to generate electricity.

The weight of more mud and sand, along with time, turns lignite into bituminous coal. When burned, bituminous coal produces twice the heat of lignite. Buried still longer, or deeper, bituminous coal became anthracite, almost pure carbon, a shiny black rock known variously as blue coal, blind coal, crow coal, stone coal, and hard coal. It burns hot, almost smokeless, with short blue flames. It produces little more heat than would come from bituminous coal, but it burns cleaner.

Marco Polo, traveling through northern China in the thirteenth century, saw coal in use. "Throughout the province" he wrote, "there is found a sort of black stone, which they dig out of the mountains, where it runs in veins. When lighted, it burns like charcoal, and retains fire much better than wood; insomuch that it may be preserved during the night, and in the morning be found still burning."

※

I fly to the Netherlands, visiting family in the south, in Noord-Brabant, an area of neat rows of asparagus and mixed crops and the smell of pigs. The Netherlands, in Dutch *Nederland*, the low country or the low-lying country, is a nation of swamps and marshes, barely above sea level, in places below sea level, dry only by virtue of pumps and floodgates. It is a haven for peat.

My father-in-law, a man of eighty years, tells me of burning peat. During the war, he says, when the Germans occupied Holland, coal was not available. His family burned peat. They burned peat millions of years too soon for it to have become coal.

I visit the Groote Peel, a park that was once at the heart of the Dutch peat-mining industry. A display shows photographs of peat diggers and their tools. The men are weathered and lean, smoking, with the thousand-yard stare of manual labor. Their tools are wooden wheelbarrows and various kinds of shovels. There is the *stikker*, with a wide, flat blade; and the *kortizer*, a very short-handled spade; and the *klotspade*, almost identical to the sharpshooter spades used by soil scientists.

With the shovels, the men dug out bricks of peat, dripping clods of roots and decaying organic matter, eight inches long by five inches wide by three inches thick, the size of a large building brick. The men stacked the peat bricks in the sun to dry, leaving behind shallow pits with square sides, never more than a few feet deep.

The Dutch author Toon Kortooms lived near here, a short bicycle ride away. "Peat is a thick book," he once wrote.

Born in 1916, Kortooms was one of fourteen children. His father managed a peat factory, where peat was converted to *strooisel,* or fuel pellets. His books put the boggy region of the southeastern Netherlands, the Dutch Peel, on the map. Among these books was the novel *Mijn Kinderen Eten Turf,* or *My Children Eat Peat,* based on the life of his father. The children, however, did not eat peat. Kortooms wrote figuratively. The father's money had come from peat, and the food for fourteen children came from the father's money.

In the book, as in life, the father burned peat in his stove. Late in the evening, he covered the burning peat with potato peels, forcing it to smolder through the night. In the morning the fire remained warm. He called the fire *de Goden*—the gods.

Rainy years were disastrous for peat mining. Dry years were better but could result in fires.

Peat pellets left the factory by steam-powered trains and wagons and long narrow boats that navigated the network of drainage canals. The pellets—*strooisel*—were used locally, but they also found their way to India, Palestine, South Africa, and the Canary Islands.

A competitor spied on the father's business. The competitor followed customers and offered lower prices but sold peat that was not quite dry or that was laced with mud. "Het leek wel modder," the competitor's customers complained, "It looks like mud." The customers returned to the father's business, to a better-quality peat, happy to pay for a superior product, happy to become part of a rambling and eventually tedious Dutch parable of success.

The book ends with the father's death. With him the era passes, the end of peat mining in the Peel. The Peel dies too, Kortooms tells us, the peat marshes killed in the peat-processing machines, converted to pellets. Where the peat was mined, only heather will grow, "small and desperate."

In the Netherlands, large-scale peat mining is finished, but elsewhere it continues. Peat is burned to dry malted barley in Scotland, giving the famous peaty flavor to the best of whiskeys. It sometimes warms drinkers in Irish pubs, burned as bricks but more often as processed peat pellets, thoroughly compressed and dried so that it burns almost without visible smoke. A power plant in Finland burns peat to generate 190 megawatts of electricity, enough to power a city.

※

A small museum honors Kortooms, the father's son, the miner of Dutch words, who died in 1999. The museum sits near the park, near the Groote Peel where his father mined, now a national park of swamps and heath and ponds that are the final product of peat mining. Here in the park, I follow a trail behind the visitor center, past a beekeeping display and its collection of hives, up to a viewing tower. From the top, I can see across the park to higher ground with fields of asparagus and mixed crops and cattle. Nettles grow along the edges of the fields. Despite the drainage canals, the park remains low lying and wet, a park as much for the sake of nature and history as for the reality of soils too saturated for farming.

Past the viewing tower, I come to a peat digging, a square-edged puddle, here as a demonstration. Twenty or thirty bricks of peat stand stacked next to the digging. The bricks are brown and laced with the stems, roots, and twigs of plants that died before I was born, but probably not too long before that, making them fibric peat, or young peat, less decomposed than hemic or sapric peat, a few million years short of becoming coal.

Next to the digging and the drying bricks, a sign, in Dutch, says, "Please do not take the peat." I pilfer two bricks, stuffing them into my pack, and leave a generous donation on my way out.

※

By car, it is three hours and a hundred pounds of carbon emissions to go from southern Holland to Assen, in the north. I am here with my fourteen-year-old son to visit a bog person. The bog person, murdered two thousand years ago and embalmed in a peat bog, lies stored under glass in Assen's tiny museum.

At the end of winding hallways and up short flights of stairs, beyond cases of stone ax heads and spear tips and hammers, beyond more cases of fired clay pots and bronze knives and a bronze tattooing needle—an inch long, green with oxidation, labeled "Bronzen tattoeernaald"—we find an isolated room. In the room, in her glass case, the bog person Yde Girl lies staring upward, eyeless, her body covered by a rough brown blanket, only her head and shoulders and feet and one hand exposed, the rest of her missing or too mangled or decomposed for display. Her skin, shriveled with desiccation, blackened by the tannins of peat, hugs the bones of her face. A lock of auburn hair lies next to her. A tiny tuft remains attached to her blackened scalp. Her mouth is open, as if in agony.

She lived around the time of Christ, around the time of Pliny the Elder. At the age of sixteen, she was killed, apparently by strangulation, possibly a victim of religious ritual. Nineteen hundred years later, in 1897, she was found by peat miners. Their tools left scars on her mummified carcass. The miners or their friends took her teeth and some of her hair and possibly a bone or two. They left intact the cord used to strangle her, which remains around her neck, under the glass, in the case.

I lean in for a closer look. I am inches from the pre-Roman Iron Age, inches from a woman who had lived at a time before matches, before steam engines and oil, before Faraday and Tyndall. Her people would have relied on wood for fuel, and possibly peat. They would have eaten bread from rough grains, the sort of bread that is available today only in boutique bakeries at premium prices.

I stare for some time, hovering inches above her glass case, examining her dry skin. I search for tattoos, a common feature in bog people, but I find none. More importantly I search for a connection to the early Iron Age, to a time that laid the foundation for the Industrial Revolution, a time when people were learning the basic secrets of manufacturing. With her mummified body in front of me, I try to imagine her in life. I try to understand how Yde may have lived and thought. I want to reach through the glass, to feel her preserved skin. I want to smell her long dead hair.

My son has but one word for my behavior toward Yde. "Creepy," he says, and he backs away.

❋

By bicycle, I ride to an Iron Age village close to the peat mines described by Kortooms. Two thousand years ago this was the site of an actual village, but today it is a living-history museum organized by a group of teachers, a reenactment of life in the time of Yde Girl, the time when people in this region were learning how to smelt iron from iron ore and to beat the hot metal into axes, knives, and jewelry.

The museum is surrounded by a wooden palisade. Inside sits a taste of Yde Girl's life—impressive huts made from wood and mud with thick roofs of thatch, a pigsty, dugout canoes, chickens and ducks and a goose, piles of firewood in ten-inch lengths. Dirt carpets the floor. There are baskets of acorns and hard white cabbages in the shadows. Along the walls are bunks with loose hay instead of mattresses and a few pieces of dirty sheepskin for blankets. The bunks surround a fire circle in the middle of the floor, an open hearth of built-up earth and small stones.

Chimneys would not come to this part of the world for at least a thousand years. The Iron Age hearth sent its smoke upward, rising to find its way outside through vents just beneath the peak of the roof. Firedogs—props of stone or clay—kept burning wood off the ground, allowing air to pass through from beneath, to feed the flames from below, and the people knew the value of dry wood. They knew that dry wood burned clean, with little smoke, and gave more heat than green wood. Soot does not stain the beams or the thatch at the top of the hut.

I try to think of Yde Girl here, probably blue eyed, her auburn hair filthy, possibly matted. She would have smelled of livestock and humanity and smoke. She may have tended sheep and goats and perhaps a cow or two. There were dogs underfoot. And small children—birth control was nonexistent, and infant mortality was high, necessitating high replacement rates. Would she have suffered from fleas? She would have worked in fields of wheat and barley and oats and various vegetables. She would have been sent into the nearby forests and meadows to find berries and edible leaves and tubers and mushrooms. She was likely tasked, on a regular basis, with grinding grain, pounding it with a heavy mortar and pestle of wood or maybe stone, or turning a grinding wheel, the beginnings of mechanization. Her hands were calloused.

The fire in the middle of her floor burned most of the time. Someone—perhaps everyone, perhaps Yde Girl—spent untold hours gathering, cutting, and splitting wood and maybe digging peat. Fuel may have been a limiting factor for the village. Too many houses, too many fires, and the nearby wood supply grew lean. As the village expanded, its inhabitants traveled farther and farther for fuel. At the end of the day, Yde Girl sat somewhere in the hut, mesmerized by the flames, feeling the warmth, watching a thin stream of smoke meander toward the openings in the roof.

Then something went wrong. She was strangled. Her body was dumped in the bog.

Outside, scattered around the village, I find more open hearths but also a clay oven, shaped like an igloo, two feet tall and three feet around. Someone like Yde Girl would have loaded the oven with wood and tended the fire while the clay walls warmed, the Iron Age equivalent of preheating. When the fire burned out, someone pushed rough dough inside and shoved a wooden door into place. The bread baked.

Nearby, a fish smoker stands, square sided with a door for fuel and wooden slats for a roof. Hanging from a stake in front of the oven: a fish, petrified by smoke, gray and dried, its eyes gone, mouth gaping, a piscine version of Yde Girl.

Near the smoker, I find a charcoal kiln. The kiln is little more than a dirt mound, eighteen inches tall and covered by grass. Fuel was stacked in a hole in the ground, sticks leaning against sticks to form a teepee, and lit. The teepee was covered with turf, the supply of oxygen choked. The wood smoldered beneath, sending out white smoke at first, then gray smoke, and finally black smoke. In a process still used today, water vapor baked out of the wood, followed by the volatiles—anything that turns to gas at temperatures close to seven hundred degrees—leaving behind blackened friable chunks of charcoal, almost 100 percent carbon, a necessary ingredient for the smelting of iron ore, or the kind of ore known as bog iron.

In wet ground near the village, a trickling spring of groundwater brought dissolved iron to the surface, where it found oxygen.

The iron combined with oxygen to become a reddish stone, an iron oxyhydroxide, a chunk of goethite. It became bog iron. It became the raw material that went into a primitive smelter.

A few steps from the kiln stands an Iron Age smelter, a cylinder of clay, its three-inch-thick walls reinforced with sticks. Bog iron and charcoal could be dumped in through an opening at the top of the kiln.

I talk to a docent in a coarse dress, a dedicated reenactor, her hair tangled and one cheek smudged with soot and her fingernails rough. She makes me think of Yde Girl. Her English falters, one word at a time, each requiring conscious effort, but it outweighs my Dutch. She describes herself as "an experiment archaeologue." She is a teacher but works here now, in part because of her love of fire, her fascination with fire. "It is what you can see in the fire," she says, "and about what you can know."

She tells me about smelting. She saw it done once. Fellow reenactors converged here from several countries, people with an interest in, and a knowledge of, primitive smelting. The little smelter's fire burned for three days and three nights. Into the smelter, the reenactors dumped bog iron. Repeatedly they dumped in charcoal. Repeatedly they added wood. Temperatures would have pushed two thousand degrees.

The charcoal produced carbon monoxide. The carbon monoxide, heated, stripped the oxygen from the bog iron to become carbon dioxide, leaving behind increasingly pure iron. The iron—droplets of liquid metal—fell to the bottom of the smelter and floated in a pool of liquefied silica sand, liquid slag, fayalite. The droplets of metal joined together. The slag, as it accumulated, was drained from a small opening in the bottom of the smelter.

After three days, the iron makers let the fire die. They broke through the side of the smelter. Inside, they found metal stuck to the clay walls. They scraped it off and took the iron to a smithing hut just behind the smelter.

The docent leads me into the smithing hut. A pair of twenty-inch-long bellows, their bags made of coarse, yellowing leather, sits on a table. In front of the bellows, a clay clamp stands ready to hold heated metal. Here the reenactors heated and pounded their iron into a useful tool. That tool—the result of three days of work, of well over a ton of carbon emissions, of the skills of a team of reenactors escaping from their twenty-first-century lives to test themselves, to apply their collective wisdom in the re-creation of Iron Age technologies—is a single spear tip, a few inches long.

Two thousand years ago, Yde Girl may have watched someone beat iron into useful shapes, sending up sparks and flashes and emitting bangs and hisses as he dropped hot iron into water, its heat suddenly quenched. Technology like this must have appeared magical to Yde Girl. The smith was almost certainly a man. As a specialist, a man who worked metal, he likely wandered from village to village, among the first of the tradesmen, likely a storyteller, by virtue of the demanding nature of his job and the anonymity of a wandering life something of a glutton compared to the men of the village.

The docent leaves, moving back to an open fire where she cooks dough balls for other visitors. When she is gone, I return to the smelter. I poke around in the ashes, hoping to find a missed bit of iron. I find a rusty nail, its presence there inexplicable, but no raw metal, nothing left over from the Iron Age, and I move on.

❋

In 1744, Benjamin Franklin worried about fuel shortages. "Wood," he wrote, "our common Fewel, which within these 100

Years might be had at every Man's Door, must now be fetch'd near 100 Miles to some Towns, and makes a very considerable Article in the Expence of Families."

Still worried four decades later, he wrote, "We must have a chimney in every room." He worried that every servant would soon want a fire. Visiting Europe, he worried that the demand for fire had consumed all the firewood in England and would "soon render fuel extremely scarce and dear in France, if the use of coals be not introduced in the latter kingdom as it has been in the former."

In the former kingdom, in England, coal had been in use for some time. To people accustomed to burning peat, the burning of soft coal would have been no more than a small step, part of a natural progression. From soft coal to hard coal would have been an even smaller step. As early as 1228, a London street was called Sacoles Lane, or Sea Coals Lane. In the early days, before mining, coal was found at the surface where veins were exposed along bluffs on the seashore and along riverbanks. It may have been called sea coal to distinguish it from charcoal, which at times was also called coal, or it may have been called sea coal because it was transported by ship to London.

But by the 1500s there were more people, all hungry for coal and the things that its heat could make. Kilns built for iron smelting deforested Kent, Surrey, and Sussex.

By necessity coal found its way from smelting furnaces to domestic use. The transition from wood to coal required chimneys. The smoke from dry wood was manageable—some even considered it aromatic. Low-grade coal, on the other hand, produced an acrid smoke that no one would consider aromatic. It was a smell that no one could love. The era of the open hearth, the era of a fire on the floor sending smoke through an opening in the eaves, died.

Within a lifetime, coal became known to common people.

From a man named Harrison, writing in 1577, lamenting the loss of the open hearth: "Now we have many chimnyes, and yet our tenderlings complaine of rewmes, catarres, and poses; then had we none but rere doses, and our heads did never ake. For as the smoke in those days was supposed to be a sufficient hardening for the timber of the house, so it was reputed a far better medicine to keep the good man and his family from the quacks and the pose."

In 1661, John Evelyn wrote his pamphlet *Fumifugium, or The inconveniencie of the aer and smoak of London dissipated together with some remedies humbly proposed by J.E. esq. to His Sacred Majestie, and to the Parliament now assembled*. In it, he compared London to Mount Aetna, the Court of Vulcan, Stromboli, and the Suburbs of Hell. Soot coated the city. It found its way inside, Evelyn wrote, "insinuating itself into our very secret Cabinets, and most precious Repositories." Raindrops fell thick and black, picking up soot and coal dust from the London air as they fell. Coal-thickened air destroyed gardens. It soiled clothes. It led to the tradition of black umbrellas.

But the Crown profited from coal. King Charles I, before the civil wars, held a monopoly on coal. The Crown taxed chimneys. If you owned a chimney—and that was the only way to burn coal—you paid.

From Lord Macaulay, writing in the 1800s, a century after the chimney tax was repealed: "The tax on chimneys was, even among direct imposts, peculiarly odious." The tax collector—the chimneyman—could barge into a home and search each room for chimneys, demanding payment of taxes on the spot. "The poorer householders," Macaulay wrote, "were frequently unable to pay their hearthmoney to the day. When this happened their furniture was distrained without mercy."

Old women cursed the chimneyman. In exchange the chimneyman hauled away whatever he could carry. "The single bed of a poor family," Macaulay wrote, "had sometimes been carried away and sold."

Macaulay related a rhyme from the time of chimney taxes, attributing it to the Pepysian Library:

> *Like plundering soldiers they'd enter the door,*
> *And make a distress on the goods of the poor.*

❋

Despite air pollution and taxes, people needed heat. They needed heat not only in their homes but in their businesses. And their need for heat drove coal mining forward. The shallow seaside diggings of the thirteenth century grew deeper. By the middle of the fourteenth century, shafts dropped vertically into the earth and widened at the bottom to form bell pits, shaped like inverted funnels. The miners dug horizontal tunnels to cut into the black veins of coal wedged between other layers of rock. At times, miners would leave pillars of coal in place to support a tunnel's ceiling. Then, as the tunnel was abandoned, they mined the pillars and made their way out.

Men, women, and children from small villages and farms who found their way to work underground developed a rough culture, a miners' way of living and talking and interacting. In the seventeenth century, they were described as swearers and drunkards, "skums and dregs" who had been driven from their rural homes, "some having towe or three wyves a peece now living." The men worked the mine face, and women and children hauled coal to the surface. From there it was carried by people and horses and carts onto ships. The ships were called coal carriers. In 1596 there were two hundred coal carriers along the coast of England. By 1615 there were four hundred. By 1635 there were six hundred.

A historical footnote: Captain James Cook, explorer extraordinaire, learned to sail on a coal carrier, a coaster that shipped black rocks to London.

A second historical footnote: both the HMS *Endeavor* and the HMS *Resolution,* the ships used by Cook on his voyages of discovery, started life as coal ships, originally named the *Earl of Pembroke* and the *Marquis of Granby.*

※

Picture yourself as a miner in the early days of coal. The shaft is dark and damp and cramped. The air you call "choke damp" can suffocate you, and the air you call "fire damp" can explode. Both are invisible. You do not know it, but choke damp is air without oxygen, and fire damp is methane. What you do know is that healthy men suddenly drop dead, overcome by choke damp, and mines suddenly explode, burning with fire damp.

You have heard of a case in which nine miners dropped dead at once, together. You met someone who nearly died, but, being hauled from the mine, he regained consciousness. He is slow-witted.

You work in a deep mine. You know that fire damp might be present. It seeps from cracks in the coal veins. Sometimes you can hear it in the darkness. Once you were knocked down in an explosion.

Your foreman has hired a fireman. The fireman is well paid. He is not the kind of fireman who extinguishes fires. His job: crawl into a mine tunnel with a candle on the end of a stick, staying low in case the fire damp ignites so as to allow the flames to pass for the most part overhead. He pushes the candle into a dead-end chamber, where the fire damp accumulates, intentionally igniting it. Today his luck holds. The fire flares suddenly, then becomes smaller, and finds its way back into a crevice.

When the fireman finishes, there is a third kind of gas to deal with, the white damp, carbon monoxide left over from the flames. So you wait. Maybe you take a caged canary with you to see if it will stay on its perch or drop to the bottom of the cage, dead. And after a while you start work. You start to mine.

In the coal you find impressions of plants, but plants unlike those you know at the surface. You suspect something dark, an underworld connection that would explain the strange plants and the sulfurous odors and the deadly gases.

To manage ventilation, flaps of canvas are draped across tunnels, or wooden doors are installed. The flaps and doors are called "traps," and they are operated by "trappers." As you approach, pushing a cart full of coal, a trapper opens the trap, lets you through, and closes the trap. The trapper is a child.

From a parliamentary commission report in 1842, the testimony of a twelve-year-old girl: "I have to trap without a light, and I'm scared. I go at four and sometimes half past three in the morning and come out at five and half-past. I never go to sleep. Sometimes I sing when I've light, but not in the dark."

Charles Dickens was once involved with investigations of child labor in coal mines. In 1841 he thought of writing an article on the topic, but if the article was written, it has since been lost.

Dickens himself, like many others, had been a child laborer. Later, writing as the narrator in *David Copperfield*: "I know enough of the world now to have lost the capacity of being much surprised by anything; but it is a matter of some surprise to me, even now, that I can have been so easily thrown away at such an age."

Like Benjamin Franklin, Dickens often published his own work. He owned the weekly literary magazine *All the Year Round*, and he was half owner of its predecessor, *Household Words*. In an 1850 edition of *Household Words* he printed an interview with a coal miner. The miner talked of being an "undergoer" tasked with crawling into holes to undermine blocks of coal, allowing the blocks to be extracted.

"In general the miner does not use the pick," Dickens was told, "and become a holer or undergoer till he is one-and-twenty. I was set to do this at nineteen, and earned four shillings a day, and sometimes more. Got badly burnt once at this work. I was lying in a new working where the air was bad, and I was obliged to use a Davy lamp. I had bought a new watch at Tipton, and I wanted to see what o'clock it was by it—else, what was the use on it?—and as I couldn't tell by the Davy, I just lifted off the top—and pheu! went the gas, and scorched my face all over, so that the skin all peeled off. It was shocking to see. I was laid up with this for two months—and sarv'd me right, I say now, but it was hard to bear at the time."

❋

Dickens did not necessarily need lamps to ignite his characters. He wrote of the spontaneous combustion of humans. Believers in spontaneous human combustion say that a person could be sitting peacefully or walking down the street when ignition occurs without warning. Once burning, the spontaneously ignited person cannot, in general, be extinguished.

In *Bleak House*, Dickens's character Krook, alone is his room, combusts spontaneously. "There is a smouldering, suffocating vapour in the room," Dickens wrote, "and a dark, greasy coating on the walls and ceiling."

The surviving characters notice a pile of white ash. They realize the ash is all that remains of Krook. "Call the death by any name your Highness will," Dickens wrote, "attribute it to whom you will, or say it might have been prevented how you will, it is the same death eternally—inborn, inbred, engendered in the corrupted humours of the vicious body itself, and that only—spontaneous combustion."

Dickens was neither the first nor the last to write of spontaneous human combustion. It is a phenomenon that is difficult to explain. In fact, it is a phenomenon that cannot be explained. It cannot be explained because it exists only in the minds of believers and the pages of writers. But it is a phenomenon that comes in handy when a novelist, trapped in a literary corner, has to kill off an inconvenient character. And who could blame Dickens, creator of so many characters, if now and again he had to dispose of one abruptly and without the entanglements of a long illness or a murder investigation?

I call the Firewalking Institute of Research and Education. I catch the instructor on his cell phone as he drives a truckload of firewood through Dallas. The truck is a sixteen-foot flatbed. The firewood is, of course, cedar.

"It is 105 degrees," he tells me. He has just finished loading—by hand, by himself—three cords of firewood. There is no need for me to ask about the intended use of this wood. He can have only one use for three cords of cedar.

"How's business?" I ask him. "Is the economy hurting firewalkers?"

"Things are really picking up," he tells me. "Things are turning around. Corporations see that. They're starting to send employees back for motivational training. They're starting to look forward. Let's get our heads out of the gutter. Sure, things were tough for a while, but business is getting better. People need to be energized."

I ask if he has ever seen peat burn. He has not. He has not even heard of peat fires.

I ask if he has seen coal burn. No again. He has never seen

a coal fire. His focus is cedar. More accurately, his focus is the heat of burning cedar and what it can do for those willing to stroll across its dancing molecules.

※

The trouble with mines—aside from cave-ins and deadly vapors and labor unrest and explosions—was their tendency to flood. Dig a hole in the ground, and eventually it fills with groundwater.

When miners tunneled into the sea or into the bottom of a lake or river or into an abandoned flooded shaft, water rushed in, fast and deadly. Near the end of June 1833, two men were fishing in the river Garnock in Scotland. They watched water disappearing into the riverbed. The flow grew stronger. The following afternoon, a hole joining the river to the underlying mines suddenly enlarged, draining the entire river. A boat was sucked into the vortex. Where there had been six feet of water, there was now mud.

More often flooding was slow and insidious, a matter of seeping groundwater or rainwater permeating down from above, through the soil. Shafts were sometimes lined with wood or brick or cast iron to seal them from seeps, yet the water would find a way into the mines. For mines dug near cliffs or bluffs, the solution was a simple drainage tunnel running downhill and out. Often drainage tunnels were narrow, no broader than a man's shoulders and four feet tall. Some of these tunnels were miles long.

As the miners dug deeper, drainage tunnels became inadequate. Chain pumps were used, with plates attached to the chain in such a way that the chain drew them up into a pipe where they would carry water upward. Egyptian wheels— a string of buckets—were also used. In both cases, power was supplied by humans and horses and donkeys or, where circumstances allowed, waterwheels fitted to streams or wind-mills.

In 1610, Sir George Selby took the problem of flooding to Parliament. Mining at Newcastle, he said, was coming to and end. Just as the smelters were becoming dependent on coal instead of charcoal, Selby predicted a coal shortage. He predicted an energy crisis.

Enter the Englishman Thomas Savery. By his own account, he drank most of a bottle of wine, then tossed the empty bottle into the fire. He watched the dregs of the wine boiling. With a gloved hand, he reached into the fire, extracted the bottle, turned it neck down, and plunged it into cold water. The loss of heat caused the gas inside to contract, creating a partial vacuum. The bottle sucked in water. In 1698, Savery patented his so-called Fire Engine, marketed as the Miner's Friend. From the patent: "A grant to Thomas Savery of the sole exercise of a new invention by him invented, for raising of water, and occasioning motion to all sorts of mill works, by the important force of fire, which will be of great use for draining mines, serving towns with water, and for the working of all sorts of mills, when they have not the benefit of water nor constant winds; to hold for 14 years; with usual clauses."

Despite the patent, Savery was not the first to apply heat to water. He was not the first to put steam to work. As early as 120 BC, not long before the time of the Yde Girl but in a more technologically advanced part of the world, Hero of Alexandria used steam to turn a hollow sphere, with the steam expelled through vents to create propulsion. In 1543 the Spaniard Blasco de Garay reportedly propelled a ship with steam-driven paddle wheels. In 1629 the Italian Giovanni Branca described what might be thought of as the predecessor to the steam turbine—a wheel with panels or vanes driven by steam. And in 1655 Edward Somerset, Marquis of Worcester, published a book de-

scribing one hundred inventions, among them "a Bridge portable in a Cart with six horses, which in a few hours may be placed
over a River half a mile broad," a device to "safely and speedily
make an approach to a Castle or Town-wall, and over the very
Ditch at Noon-day," and a steam engine almost identical to the
one Savery patented forty-three years later.

And from Samuel Morland, writing in 1683: "Water being
converted into vapour by the force of fire, these vapours shortly
require a greater space (about 2,000 times) than the water before occupied, and sooner than be constantly confined would
split a piece of cannon. But being duly regulated according to
the rules of statics, and by science reduced to measure, weight,
and balance, then they bear their load peaceably (like good
horses), and thus become of great use to mankind, particularly
for raising water, according to the following table, which shows
the number of pounds that may be raised 1,800 times per hour
to a height of six inches by cylinders half filled with water."
What he meant was that steam was powerful but could be controlled, like good horses.

Savery, like Morland, thought of power in terms of horses,
of horsepower. And he knew that horses needed breaks. They
needed time to sleep. They needed time to recover from injuries.
To get two horsepower all day long, day after day, required as
many as ten or twelve horses, with two working while the others
rested. "So that an engine which will raise as much water as two
horses," he wrote, "there must be constantly kept ten or twelve
horses for doing the same."

※

Despite all the earlier work, despite Savery's patent, steam as a
mode of power did not take off right way. It took Thomas Newcomen's engine, the atmospheric engine built in 1712, to catch
on. The atmospheric engine was the first steam engine to put

horses out of work. Steam from a boiler filled a cylinder, lifting a piston upward. The volume of water as steam was two thousand times that of water as liquid. The cylinder full of steam was cooled, condensing the steam into water and creating a partial vacuum. Air pressure—atmospheric pressure—pushed the piston down. Steam powered each upward stroke of the piston, and atmospheric pressure powered each downward stroke.

Newcomen's atmospheric engine was a fuel hog, but since its main use was in dewatering mines, coal was readily at hand. This reality stalled improvements.

But James Watt recognized the inefficiency of the atmospheric engine. He saw that repeatedly cooling and reheating the steam in the cylinder accounted for three-quarters of the fuel use. Although the miners already owned atmospheric engines, Watt moved ahead, designing an engine with a separate condenser—a vessel attached to the piston in which the steam could be cooled and condensed without cooling the piston itself, doing away with the repeated cooling and reheating of the atmospheric engine's cylinder. When Watt completed his first engine in 1775, it included other improvements, but the key was the separate condenser and its fuel savings.

Watt's engine was to the atmospheric engine as a hybrid car is to an SUV. But so what if Watt's engine used less coal? With the dewatering power of the atmospheric engine, there was plenty of coal. Why upgrade when fuel was cheap? From Dionysius Lardner and James Renwick, in their 1856 book *The Steam Engine Familiarly Explained*: "Notwithstanding the manifest superiority of these engines over the old atmospheric engines; yet such were the influence of prejudice and the dislike of what is new, that Watt found great difficulties in getting them into general use."

Watt and his business partner Matthew Boulton developed a marketing scheme. They would supply their engine to miners, and in exchange the miners would pay a licensing fee computed as a portion of fuel savings. It was the equivalent of leasing a hybrid car in exchange for a portion of the savings on gasoline.

Watt and Boulton accumulated a fortune. And because their engine needed less coal, it could be used to provide power well removed from mines. It could be used to power mills and factories. Watt and Boulton had harnessed heat. They came up with an economical method of turning coal into work.

The steam engine moved from stationary duties to mobile duties, to ships and trains. George Stephenson, a miner from a miner's family, raised in the poverty of mining, came across one of Watt's engines in Scotland. In 1813, with the help of a blacksmith, Stephenson built his first engine. This was at a time when the mines still relied on horses to haul coal from the tunnels. This was the time of the early steam-driven carts and suggestions that wheels should be fitted with hooflike spikes to ensure traction. It was also a time of serious concerns regarding the usefulness of trains for passengers. It was a time when futurists conceptualized passenger trains but wondered if they would be useful, if enough people would regularly want to go to the same place at the same time.

Stephenson's *Blücher,* his first train engine, went to work in 1814, hauling thirty tons of coal uphill at the pace of a brisk walk. He built more engines. In 1829, with his son, he built the *Rocket.* The *Rocket* was entered into a contest. The contest was held to find an engine that could be put into general use, that could be supported by investors seeking to change the world and in so doing make a bit of money. The *Rocket* won, hands down, covering thirty miles in two hours, fourteen minutes, and eight seconds, at one point hitting twenty-nine miles per hour. Its secret: heat from the coal fire was distributed to the boiler through twenty-five tubes, each three inches in diameter, winding back and forth through the firebox, drawing as much heat as possible as quickly as possible into the water inside the tubes. The *Rocket* captured heat as no other engine had before.

Watt and Stephenson changed the world. Or, closer to the point, Watt and Stephenson contributed to the changing world, but their names took hold, became iconic, became part of the technological revolution that would, over the next century and a half, pump three hundred billion tons of carbon out of the earth and into the air.

Ralph Waldo Emerson died in 1882, five decades after the *Rocket*'s twenty-nine-miles-per-hour speed record, fourteen years after Tyndall wrote his biography of Faraday, and fifteen years before Yde Girl's mummified remains were discovered by peat miners in the Netherlands. Before he died, Emerson wrote about coal. "For coal is a portable climate," he wrote. "It carries the heat of the tropics to Labrador and the polar circle; and it is the means of transporting itself whithersoever it is wanted. Watt and Stephenson whispered in the ear of mankind their secret, that a half-ounce of coal will draw two tons a mile, and coal carries coal, by rail and by boat, to make Canada as warm as Calcutta; and with its comfort brings its industrial power."

With my family, I drive sixty miles due east, crossing into Germany, to Essen, the site of the Ruhr Museum, a world heritage site that started life as a coal-washing plant, part of the Zollverein coal mine, shaft number 12. Some of the docents are retired miners.

Carboniferous coal—coal from three hundred million years ago—underlies the whole region in layers as thick as fifteen feet.

❋

Mining here started in 1847 with a shaft five hundred feet deep. The shaft flooded and was abandoned. Another shaft was dug. And a third. Pumps were introduced. By World War II, two thousand workers lived here, miners and plant personnel and engineers and bookkeepers. When the mine closed in 1986, the shafts had passed three thousand feet in depth, deep enough to feel the warmth of the earth, 104 degrees at the mine face. The miners here experienced heat cramps, dehydration, and occasionally heatstroke.

The entrance to the museum—the entrance to the washing plant—is through the shaker room, with black and dark red pipes and steel chutes and giant gears and shaker trays where larger coal was separated from smaller coal and from dust.

Coal came from the ground and into the plant on black conveyor belts. The belts remain intact, part of the museum. Rusting small-gauge railway tracks run through the building. Wooden platform wagons, five feet long and three feet wide, sit in corners, their purpose forgotten. Now the machinery lies idle and rusting, an industrial tomb turned museum. In life, the motion was unceasing, the noise overwhelming, rocks tumbling against rocks or sliding down steel chutes, rattling conveyor belts, grinding gears, beating engines. Black dust filled the air— dust full of solar energy captured by plants that became peat that became coal that became a portable climate and industrial power.

The Iron Age smelter, the Iron Age blacksmith, transported here when the plant was alive, would have been terrified. Likewise, the average modern office worker, transported here when the plant was alive, would be terrified.

In addition to sorting and cleaning, some of the coal was baked to become coke, using the same process that converts wood to charcoal, heating in the absence or near absence of oxygen. The coking process is a cooking process, the two words related. Coking, or cooking, drives off moisture and volatiles. Coking cooks out the smells. The coal melts, becoming plastic and then tar. It migrates toward the edges of the oven. It contracts and moves back toward the middle of the oven. Around 1,500 degrees, the coke has been cooked into a glowing lump of almost pure carbon. Coke burns without the odor of coal, and in blast furnaces coke provides the carbon needed for smelting, the carbon needed in the chemical reactions that separate the iron from the ore.

Most signs are in German, but a few are translated into English: "Fire and the underground world can be termed as the original myths of the Industrial Revolution," and "the raw call" in front of a display case with raw coal, and "slack" in front of a piece of slag, a piece of coal ash, the burned-out remains of a tree that lived millions of years ago.

Electronic billboards display current unemployment numbers for the Ruhr. The valley of the Ruhr, it turns out, is no place to find a job. It was worse after World War II. Coal production suddenly dropped 90 percent. The mines themselves were damaged by the war, but more damaging was the unavailability of food and housing. Equipment suffered from neglect. Bridges had been bombed, railroads destroyed.

From Chauncy Harris, writing in 1946: "Most of the homes of Europe have been cold this winter. They lacked the coal needed for heating. Most of the factories have lain idle. They lacked coal for power. The coldness, idleness, and darkness of Europe are due to many causes, but the most important single factor is the breakdown of coal production in the vital Ruhr coal-mining district of northwest Germany."

Glass cases hold curiosities that came up from the coal mines. One holds a white and brown tube of silica and quartz with bubbly edges. It is three feet long and maybe an inch wide, and eighty million years old. It is a *Blitzröhre*, a thunder tube, but in English a fulgurite, a piece of ground struck by lightning, three hundred kilowatt-hours delivered in an instant. There are fossils: mussels from 300 million years ago, an ammonite snail from 100 million years ago, a whale bone from 26 million years ago, fern fossils from 315 million years ago, an undated chunk of lava.

My best find is an equisetum stalk, a roadside weed, common along marsh edges, also common as a fossil in the abandoned mine cuts north of Anchorage, often called horsetail, which today grows to the size of a dandelion. It is a primitive plant, a nonflowering relative of the more advanced plants, evolutionarily closer to a fern than a pine tree. But unlike today's horsetail, the one in this museum is five inches in diameter and six feet tall, a small tree. Three hundred million years ago, it would have been possible to walk through a forest of equisetum trees and tree ferns with canopies of club moss. It would have been possible to touch the related lepidodendron tree, ten stories tall, and to feel its alligator-skin bark. But there were no people.

My son, bored, leads me to the cafeteria. We eat overpriced bratwurst. On the table, a flowerpot holds water-soaked coal and two sickly flowers, someone's idea of a decorative accent. I pocket a piece of coal, black and shiny, a half ounce in weight. And another. Two Germans look on in wonder, or perhaps disapproval, their mouths bulging with bratwurst.

We retreat up a flight of stairs to emerge outside, on top of the plant. The view is one of pipes and block-shaped buildings and silver tanks and pulleys the size of Ferris wheels with cables that run underground, interspersed with thick stands of trees. Among them we see houses, green and yellow and white, and apartment blocks with roofs of red and brown. Three church spires, built well before the mines, stand above the trees. More may be hidden in the smog.

I count nineteen smokestacks. They stand higher than the church spires. Most appear to be relics, disused and dead. But some, painted white, emit billowing clouds of steam. One, in the distance, emits a yellow flare. Together they emit the smell of industrious burnings, of dark, sulfurous vapors that have been locked underground for three hundred million years, the aroma of a portable climate.

James Watt, making his fortune from licensing fees based on coal saved when his engines replaced the atmospheric engines that Newcomen developed, occasionally ran into miners who powered their pumps with horses. Watt made calculations. Like Savery and Morland before him, he compared the power of his engines to the power of horses. Watt's horse, harnessed to a crank and walking in a circle, could lift thirty-three thousand pounds to a height of one foot in one minute. A steam engine could do the work of one horse for twenty-four hours by burning less than two hundred pounds of coal.

Coal was cheap. Cheap coal and Watt sent horses out to pasture.

Ten typical human workers can do the work of one horse. A human runs, sustainably, at about one-tenth of a horsepower. Two hundred pounds of coal would do the work of thirty humans for a day, assuming that a human is good for an eight-hour workday, while the steam engine runs nonstop, never sleeping, never eating, never complaining.

In 1889, seventy years after Watt's death, during the Second Congress of the British Association for the Advancement of Science, Watt was recognized. A unit of power was named after him. It was defined as one joule per second, or about a quarter of a calorie per second. One horsepower equated to about 750 watts. Watt, alive, harnessed to a crank and walking in a circle, could provide seventy-five watts of power without coal.

※

I have business in London on the way home that necessitates a day of waiting around. The newspapers talk of record heat again, of heat waves throughout western Europe, of 125 degrees in Saudi Arabia and 128 degrees in Pakistan. Smoke is drifting into Moscow, coming from peat fires that exist because swamps outside the city were drained by miners digging peat used to generate electricity. Customers are running their air conditioners in Virginia, using the heat from coal to cool their homes. About 40 percent of Virginia's electricity comes from coal-fired power plants.

I take a train to Kew Gardens station and walk to the Kew Bridge Steam Museum. The museum boasts running steam pumps and a working steam train. "Marvel at the world's largest collection of steam pumping engines," the brochure says, "many of which you can see working every weekend."

I arrive sweating. The museum sits in a brick building left over from the nineteenth century. Near the front door—near what would have been the front door before this place became a museum—a sign says that Charles Dickens visited the plant in April 1850. He passed through this rusticated doorway to marvel over the engines then in use, engines now on display, that pumped water all over Dickens's London, water for drinking and baths and industry.

I walk into the Number One Boiler House, where Lancashire boilers used coal to generate steam until they were shut down in 1942. The boilers, long since removed, would have stood in rows, each a cylinder lying down, each with a diameter as tall as a man. In front of each, someone would have stood shoveling coal in through doors near the bottom. The doors opened into chambers lined with tubes that ran the length of the boilers. Through the tubes, water circulated, boiling, generating steam.

It is possible to imagine Dickens standing here amid the noise and the heat, watching the men shoveling coal, noting their language, their posture, their clothes. It is possible to imagine Dickens converting living men into literary characters.

Dickens knew the smell of London air. From his novel *Hard Times*: "It was a town of machinery and tall chimneys, out of which interminable serpents of smoke trailed themselves for ever and ever, and never got uncoiled."

In this building where Dickens once walked, a great collection of steam engines has accumulated. There is an Easton and Amos engine from 1863, with a flywheel sixteen feet across, tastefully painted in cream and brown. There is a Dancer's End engine built in 1867, with two cylinders, each one four feet tall and two feet across. There is a Waddon engine, built in 1911, its cylinders lying on their sides and painted black.

A placard tells me that the Waddon engine could generate 198 horsepower. It pushed three million gallons of water each day into a water tank that stood 282 feet tall. This little engine, with steam and ingenuity and coal, could do the work of two thousand men.

Signs hang on each of the machines: "This engine will not be working until 4th July due to scheduled maintenance." On steam valves, another sign: "Equipment out of action. Do not use."

I find a docent sitting in a quiet corner. He tells me that even the steam train is out of commission, not running. He tells me that I should come back another time.

All the engines here are variations of Watt's machine. I ask where I might find a Newcomen engine, an atmospheric engine. The docent does not move from his chair. They have no Newcomen engines here. "They didn't run very well," he says. "Not very efficient, you see?"

I find more engines. The largest, built in 1847, just three years before Dickens came to visit, had a piston that would travel upward eleven feet on each stroke, eight strokes per minute. It sits in an open room with black iron stairs leading to the top of the cylinder. I climb the stairs and lean dangerously far over a guardrail to touch the piston itself. The piston is shining steel, oily, as big around as an elephant's leg. Another flight of stairs takes me to the Beam Room, where the engine's rocker beam transferred its power to a pump. The floor here is timber, worn in places from engineers walking back and forth, perhaps worn by the boots of Dickens himself.

I wander out into the yard, looking for boilers and stockpiles of coal. Instead I find the steam train locked in a shed of rusting sheet metal. I can see the engine through an opening in the shed's doors. It is not a full-sized engine but a small-gauge train, for tourists, painted green, a toy for steam lovers.

I go back to the docent, still perched on his chair, unmoving. I ask where the new boilers are, and the coal stockpiles. "We don't burn coal anymore," he says. "The boilers run on gas now. The city will not allow the burning of coal." The boiler room is not part of the exhibit. It is off-limits.

"What about the train?" I ask. He assures me that the train, when it runs, runs on coal. The coal, he tells me, comes from Poland.

Back in Anchorage, I burn wood in a metal fire pit behind my house. I build a bed of red-hot coals. On top of this, to the right, I place my pilfered brick of peat. To the left, I place my pilfered coal. With my infrared thermometer, I shoot a reading. The red-hot coals of burned wood, now flameless, hit a thousand degrees. The tops of the block of pilfered peat and the pilfered coal remain, for now, cool and lifeless, at sixty-five degrees.

Within a minute the peat flares, first at one end, then the other, but the flames disappear as quickly as they appear, leaving behind a red glow that consumes individual fibers. With my bellows, I blow air across the peat. The fibers ignite. In their flames, I see strata in the peat, layers of long-dead vegetation, each layer perhaps a year of growth. I count fifteen layers in the flames.

The temperature of the flames: 1,100 degrees.

The coal remains quiescent, holding on to its carbon. I encourage it with the bellows. It begins to glow. As it warms, jets of flame ignite around it, flares of vapor. I encourage it further, pumping air across it. At two minutes, it develops a steady red glow. It sends out more flares. I continue pumping. At three minutes, it produces a steady flame. It becomes a rock on fire.

The temperature of the burning rock: 1,200 degrees.

I add wood to the fire pit, separate from the peat and separate from the coal. The wood flares up within seconds. I have before me fires of wood and peat and coal. The important difference between these fuels is not so much the temperature of their flames. That temperature changes by the second, reacting to changes in wind and moisture and any number of realities. The important difference is in the amount of heat they put out per weight burned. Coal—even low-quality coal, lignite, one step removed from peat—wins. A pound of coal produces twice the heat of a pound of wood. Wood and peat, on the other hand, match each other. Peat, dry, is not much different from wood. But at times, and in places, peat was abundant when wood was not.

Even this single tiny lump of burning coal noticeably taints the aroma of wood and peat smoke. My dog, who loves company, sniffs the air, stands up, stretches, and goes inside alone, tail down and ears tight against his head. I bend over, closing in on the coal to smell its raw fumes, to inhale acrid smoke. It smells unpleasant in the extreme, not the sulfur of rotting eggs but something more sinister, a smell that assaults the nostrils, reminiscent of hot tar.

If a smell can be bitter, the smell is bitter. If a smell can be black, the smell is black. It is the smell of nineteenth-century London. It is a smell that would have been familiar to James Watt and George Stephenson and Ralph Waldo Emerson and Charles Dickens. It was, at one time, the smell of progress.

Chapter 5

ROCK OIL

In the 1850s, people burned wood and peat and coal for heat, but they had more choices for lighting. Most people burned candles, but they also used oil-burning lamps. The wealthy burned sperm whale oil, the less wealthy burned the brown oil from bowhead whales, and the even less wealthy burned the fat from pigs. For those willing to risk explosion and fire, there was light from lanterns fueled by camphene, which was distilled from turpentine, which came from pine sap. Some cities used town gas, a mix of hydrogen, methane, and the various other hydrocarbon gases that came from coal and was piped into street lamps and homes. By the 1850s, people occasionally burned kerosene in lamps. At that time, kerosene, like town gas, was refined from coal. Kerosene lamps had certain advantages, but kerosene from coal was expensive.

In the early nineteenth century, oil from seeps in western Pennsylvania, in a region that was home to the Seneca Indians, was soaked up in rags or skimmed from creeks that flowed near the seeps. Rural entrepreneurs dug pits and trenches to increase their take. In this way, a day's labor could yield a few gallons. The fluid came from the earth and was known generically as rock oil or Seneca Oil.

At first, rock oil was not burned. It was sold as a cure for toothaches, headaches, stomach pains, worms, rheumatism, and deafness. The limited medicinal properties it possessed were exaggerated to the point of hyperbole. From an advertisement for Seneca Oil:

> *The healthful balm, from nature's secret spring,*
> *the bloom of health and life to man will bring;*
> *As from her depths the magic fluid flows,*
> *To calm our sufferings and assuage our woes.*

In the early 1850s, a man named George Bissell, an entrepreneur by nature and economic necessity, acquired a quantity of the healthful balm and realized that it might be burned as an illuminant. He pulled together investors and hired one Professor Silliman to write a report on rock oil. Silliman's experiments occasionally caught fire, but he reported "success in the use of distillate product of Rock Oil as an illuminator" and described refining methods.

"It appears to me," Professor Silliman wrote, "that there is much ground for encouragement in the belief that your Company have in their possession a raw material from which, by simple and not expensive process, they may manufacture very valuable products."

Professor Silliman understated the facts. What the company had in its possession was a substance that would change the world. The molecules of petroleum carried energy efficiently and conveniently. But to change the world and to enrich himself and his investors, Bissell needed to find the stuff in quantity. He needed to get more rock oil out of the ground.

Bissell and his investors hired Edwin Drake, formerly a conductor for the New York and New Haven Railroad, a man whose chief qualifications were his amiable nature and the fact that he was unemployed. He stood tall, dressed well, and had large eyes that often conveyed a sense of deep thought and intense interest. A portrait suggests that he was heterochromic, his left eye darker than his right. Bissell dubbed him "Colonel Drake" to add an air of respectability and sent him to what was then the hinterlands of Pennsylvania, to a place called Titusville. It was May 1858, and Titusville was a logging community near the oil seeps, population 125. After having visited Titusville, one of Bissell's investors called it "the most forsaken place I was ever in." The investor commented, too, on the mud, which would, after Drake's success, become a popular theme in written accounts.

With his wife and child, Drake moved into a Titusville hotel. The hotel was known to the locals as "the tavern."

"I found it a rough place for a residence," Drake later wrote, "but with my family around me I was happy."

Drake hired men who worked the creeks and repaired existing trenches and pits. For their trouble, they harvested a few barrels in a season. Soaking oil into rags or skimming it from the surfaces of creeks was no way to get rich.

Someone—probably Bissell, but possibly Drake—remembered an advertisement for medicinal Seneca Oil, specifically for the bottled medicine sold as Kier's Petroleum. The advertisement showed drilling rigs of the kind used to drill for salt, an ancient practice that had been adopted in the salt-rich lands of western Pennsylvania. Kier's Petroleum came up with brine from wells near Titusville. Kier was for the most part after the brine, from which he harvested salt. He and the other salt drillers usually threw away the oil, dumping it into pits or rivers. But Kier's wife was treated for tuberculosis in 1848 with a prescription for American Medicinal Oil, which came from a salt well in Kentucky, and Kier recognized the stuff in the bottle. He started bottling medicinal oil from his own wells. For a decade before Drake arrived in Titusville, Kier had been saving some of his oil and bottling it as medicine, to be taken externally, as a liniment, or internally. Before Drake arrived in Titusville, Kier had sold almost 240,000 bottles of rock oil in half-pint medicine bottles.

Bissell and Drake, influenced by Kier's advertisement with its sketch of the salt wells, decided to drill. They decided to drill for rock oil—not for salt with rock oil as a by-product but for the oil itself.

Drake had trouble finding a qualified driller. The problem was not an absence of drillers but an absence of drillers willing to work for Drake. In western Pennsylvania, drillers were too busy drilling for salt. Several of them told Drake that they would come to Titusville to drill but never showed up. At least one driller promised to come to Titusville for no other reason than to get rid of Drake, to get Drake to leave him alone. The driller thought that anyone intentionally drilling for oil must be a nut. Oil was not a valuable commodity. Oil was a nuisance. It ruined salt wells.

The drillers, Drake wrote, "were thirsty souls and preferred Whiskey to any other liquid for a steady drink."

Drake persevered. He was later remembered as quiet and reserved, "never boasting or bragging about the great things he expected to do." He seldom swore. He seldom drank. Money was a problem. The investors, initially enthusiastic, did not always come through with funds for legitimate expenses. Drake was not fully paid his agreed-on salary. But his reputation for persistence grew. A local businessman loaned him money on little more than his word, and Drake got by.

Still searching for a driller, Drake hired men to dig into the earth, to dig a well. The well they dug flooded.

145

Drake acquired a steam engine, paying for it in installments. He built an engine house. But still he was without a driller.

In April 1859, Drake heard of another driller. The driller was working in the town of Salina, near Tarentum. The driller's name was William Andrew Smith, but he was known as Uncle Billy. He was not purely a driller. He was a man who could fix salt well problems, but at the time he was working as a blacksmith and planning to reinvent himself as a farmer. Instead he went to work for Drake. Their agreed-on salary: $2.50 per day, with Smith's sixteen-year-old son thrown in gratis. Smith's wife refused to go. Titusville, to the cosmopolitans of Salina, was the backwoods. In need of a cook and housekeeper, Smith turned to his daughter Margaret, age twenty-four.

"When I first saw Colonel Drake," Margaret later wrote, "he had on a black suit and a high black hat. He was naturally a dark-complexioned man. And the black hat and the black clothes made him look all the darker. After he had gone I remember saying to my father, 'That man looks like an Indian.'"

Margaret worried out loud that they might all be killed.

But instead of being killed, they moved into the engine house that Drake had built. The frogs kept Margaret awake at night. Drake and Smith raised a derrick. The derrick, like the engine house, was made from rough lumber, unpainted. It was twelve feet on each side at its base, tapering to three feet on each side at its top. It was built lying down, on its side. With the help of sawmill workers and the scattered neighbors of Titusville, Drake and Smith tipped the derrick upright.

Smith's wife and an additional four children joined him in July. Before it was over, all of Smith's children would work on the well. The method they used was not modern-day rotary drilling. They were not drilling in the same sense that one would drill a hole in a plank. They were drilling in the sense of drilling with a nail—of banging a nail into a plank, then removing the nail to leave behind a hole. They were pounding and chipping, using a method variously known as cable drilling or cable-tool drilling or percussion drilling or cable-tool percussion drilling.

They started with a pit. The pit was about eight feet deep, reinforced by logs. In the bottom of this pit, they drilled. A rope dangling from the top of the derrick held the drill bit, which was in fact a heavy chisel. Smith had made the bit—the chisel—with his own hands. The principle was simple: lift the bit and let it drop, lift the bit and let it drop, lift the bit and let it drop. The steam engine's power—power derived from burning wood—lifted the bit. Gravity drove the downward drop. Each time the bit dropped, the well grew deeper, a sliver of earth chipped away. Periodically, the bit was lifted clear of the hole, and a scoop was lowered to remove chipped rock.

At sixteen feet, they struck water.

Smith rigged a pump, but knee-deep water in the pit slowed progress. The pit collapsed. Someone—maybe Drake, maybe Smith, maybe someone else—suggested driving a pipe into the existing hole to protect it from cave-ins and to slow down the water. The method had been used before the Drake well, but the idea seems to have been new to the men in attendance. They lined their sixteen feet of bore hole with two lengths of cast-iron pipe. Drake acquired additional pipe. They drove the pipe into the ground. At a depth of forty-nine feet and eight inches, they struck rock. The bore hole was now lined with pipe all the way to a layer of solid rock, so it would neither collapse nor flood.

Through the lining of cast iron, they drilled deeper. Lift the bit and let it drop, lift the bit and let it drop, lift the bit and let it drop. Eight hundred and fifty pounds of drilling gear went up and down, pounding and chipping through the pipe, through the cast-iron casing, cutting its way toward the fat of the land.

By this point, Drake had more or less stopped paying his bills. In late August, Drake's investors sent him a letter telling him to abandon the well, to close up shop. They gave him five hundred dollars to cover the costs of shutting down. But in 1859 the mail to Titusville moved no faster than a horse could walk. Drake and Smith kept lifting and dropping, lifting and dropping.

On Saturday, August 27, 1859, they reached a depth of sixty-nine feet. Without warning, the drill bit dropped six inches. It had hit a void in the rock, a soft spot. They called it a day.

The next day, the Sabbath, a day of rest, Uncle Billy Smith realized that the void was full of oil. They had struck oil in the bottom of Drake's well.

Smith replaced the drilling gear with a pump. The pump, in the bottom of the well, was connected to the surface by iron sucker rods. The sucker rods were connected to the walking beam, which was connected to the steam engine. The exciting business of exploring for oil was replaced by the dull routine of production, of pumping, the steady chugging of the engine and the splashing of the oil at a rate, at first, of ten barrels a day.

Derricks sprung up around Titusville, a new kind of forest. Kerosene distilled from rock oil became rather suddenly available in large quantities, and the kerosene lantern became an inexpensive source of high-quality light.

By the time I arrive in Titusville, I am 150 years too late for the boom. But I am here for the historical ambience, not for a job. I check into a hotel that was once a railroad car and still sits on tracks that once carried oil to markets. That oil was burned in lamps, and later in ships, and still later in automobiles.

The tracks run beside Oil Creek, a shallow river with a rocky bed thirty feet across from bank to bank, a gurgling stream of the sort that trout call home. From 1785, seven decades before Drake, in the memoirs of General Benjamin Lincoln: "In the northern parts of Pennsylvania, there is a creek called Oil

Creek, which empties itself into the Allegheny river, issuing from a spring, on the top of which floats an oil, similar to what is called Barbados tar, and from which may be collected, by one man, several gallons in a day."

The drilling stopped the oil seeps that leaked into the creek. The seeps no longer flow. Now the creek runs clear, its surface untainted by Barbados tar.

Across Oil Creek from my hotel, the site of Drake's well has been converted into a museum. Drake's derrick has been reconstructed on the grounds of the museum. The derrick, shrouded in rough wooden planks, stands three stories tall. It is attached to an engine house, also of rough wooden planks. In that engine house, a five-hundred-gallon boiler heated by burning wood supplies steam to a six-and-one-half-horsepower steam engine. The steam engine's single piston moves a walking beam up and down, like a seesaw, and the end of that beam lifts and drops a rod that extends down into the well. Each time Drake's walking beam moved up, it lifted seventy feet of steel rods. At the bottom of those steel rods sat a pool of oil, and each stroke of the steam engine sent a spurt of oil to the surface, where it was directed into a wooden cistern.

Today, on the museum grounds, the Drake well house recirculates oil to show how things would have looked after the well was drilled—after the discovery, when the well was in production. The oil does not come from a depth of seventy feet, but it is real oil. It is oil that the museum recirculates for the viewing public. It is crude oil, smelling sulfurous and bituminous, flowing darkly, looking black at first but then, in the right light, casting a greenish tint. It flows like a thin syrup, watery, only hinting of viscosity.

From Professor Silliman's 1854 report on a sample of oil from Oil Creek: "The Crude oil, as it is gathered on your lands, has a dark brown color, which, by reflected light, is greenish or bluish.

It is thick even in warm weather—about as thick as thin molasses. In very cold weather it is somewhat more stiff, but can always be poured from a bottle even at 15 degrees below zero. Its odor is strong and peculiar, and recalls to those who are familiar with it, the smell of Bitumen and Naphtha."

The well itself—the hole in the ground on top of which the derrick sits—is as real as the oil being pumped. It is the Drake well, the hole that Drake made, still intact. With a diameter of about five inches, it is a hole that changed history.

I stand here in Drake's reconstructed engine house with its attached derrick, rigged now as it would have been then, during the pumping phase of the endeavor. The walking beam goes up and down. I listen to the steam engine's pistons, its valves popping and hissing with the regular slow heartbeat of steam-powered motion. Recirculated oil pulses out of the well and into a wooden storage tank, showing visitors what Drake might have seen after the well was drilled.

Almost everything here is a reconstruction: the engine house, the derrick, the sucker rods, the drilling tools hanging on the wall, the pump, the barrels. But the well itself is the original. The actual Drake well exists intact. I crouch and touch the metal casing going into the soil. I feel the ground. I smell the oil. This is where it started—the petroleum revolution, the displacement of coal and the replacement of whale oil, the acceleration of climate change.

But only the hole itself is the original. The rest of it burned on October 7, little more than a month after Smith's drill bit found an oil-filled void in the rocky ground of western Pennsylvania.

In an 1880 edition of *The Titusville Weekly Herald*, Uncle Billy Smith remembered the fire. It was late, about ten o'clock. Drake himself was away, off buying pipe. "I thought the tank was not filling up fast enough," Smith told the reporter, "and went to see if the oil had stopped. I had a little lamp in my left hand. A little streak of light, like a flash of lightning, went to the oil, set everything in a blaze and in two seconds everything was burned up, including the little house I lived in."

When Drake was told of the fire, he asked about casualties. Was anyone killed? Was anyone hurt? Remarkably, no on both counts. "I'm glad of that," Drake responded. Only then did he ask about the well. The works were gone—the derrick, the engine house, the ropes, the steam engine, all in ruins.

"Did the hole burn?" Drake asked. No, the hole was intact.

Without missing a beat, Drake was ready to rebuild.

The fire was the first of many well fires in western Pennsylvania and among the least tragic.

Drake was far from the first to discover oil. Pliny the Elder wrote of it in the first volume of his *Natural History*. "In Samosata, a city of Commagene," he reported, "there is a pool which discharges an inflammable mud, called Maltha. It adheres to every solid body which it touches, and moreover, when touched, it follows you, if you attempt to escape from it. By means of it the people defended their walls against Lucullus, and the soldiers were burned in their armor. It is even set on fire in water." In the next paragraph he writes of Naphtha. "Naphtha is a substance of a similar nature (it is so called about Babylon, and in the territory of the Astaceni, in Parthia), flowing like liquid bitumen. It has a great affinity to fire, which instantly darts on it wherever it is seen."

In the next paragraph, he writes of "places which are always burning," referring now not to petroleum but to volcanoes.

Drake was not the first to discover oil, and he may not have been the first to drill for oil, even ignoring the dozens of salt wells that unintentionally struck oil. A retired medical doctor named Dr. Robert Hazlett bought land along Oil Spring Run in West Virginia in January 1859. With his investors, he drilled. He did not drill for salt. From *The Derrick's Hand-Book of Petroleum*, 1898: "It is claimed that this company—known as the Virginia Petroleum Company, drilled their first producing well as early as—possibly before—Drake drilled his historic well in Pennsylvania."

The field Hazlett discovered was named the Volcano Field.

But first or not, Drake catalyzed a boom. Merchants and preachers and lawyers abandoned stores and pulpits and offices to become oilmen. Whalers, tired of long voyages and uncertain profits, set a course for Titusville. Farmers beat their plows into drill bits.

The market, for now, was lighting—the heat from burning oil, like the heat from burning whale oil or wood or coal, created light. It was the heat that made the flame incandesce. But it would be some time before the heat of burning rock oil would be converted to a mode of motion.

Downstream from Drake's well, almost to its confluence with the Allegheny River, a distance of more than ten miles, Oil Creek and the surrounding land are now a state park. I had planned to canoe the creek, to follow the route of the oil that once traveled here. But just now this creek will not float a canoe. It is too shallow. The cobbles and pebbles that line its bed reach within inches of the surface.

Even in Drake's time, the creek's water levels were known for their responsiveness to rain. Not wanting to ship oil on a schedule dictated by weather, enterprising transporters devised a series of small dams along the creek's tributaries. They opened the dams in a synchronized fashion, sending water downstream, a flash flood made to order to carry shallow-draft scows to the

Allegheny River, at the time a river of commerce. Those scows, roughly built, carried barrels of oil. They called the flash floods "freshets."

The barrels are gone. The scows are gone. The dams are gone. There will be no freshets today. So I walk along the creek's banks in on-and-off rain. The hills around Oil Creek, originally covered by conifers and then laid bare by loggers and oilmen, stand covered with mixed deciduous trees, the sort of regrowth that passes for nature in the eastern United States. Mosquitoes abound. A red-winged blackbird displays. The creek that once carried oil now carries trout. I watch a doe wading. She watches me, her brown eyes large and her ears twitching, and then she disappears into thick brush.

Along Oil Creek I find, without trying, abandoned well stems, vertical pipes standing a few inches above the soil surface, overgrown with weeds. Some of the wells stand in the company of rusting steel rocking donkeys, the remains of oil pumps. I find an old wooden storage tank that at one time may have held a hundred gallons of crude, now empty and grown over by a white-flowered briar. A four-inch pipeline runs next to my path, for the most part hidden by leaf litter and undergrowth, occasionally half buried, its outermost skin reddish brown with rust.

I stroll through Funkville. That is, I stroll past a sign in the forest proclaiming the site to be the abandoned community of Funkville, where Captain A. B. Funk drilled into the third sand. Drake's well struck oil in the void that became known as the first sand, but there was more oil to be found deeper, in the second sand, and still deeper in the third sand.

Captain Funk's well flowed without pumping, an artesian well of oil sending three hundred barrels a day to the surface. He called his well "the Fountain." Funkvillians once strolled here next to the Fountain.

After Drake's discovery, wells were routinely named. There

was the Noble, the Empire, the Craft, the Coquette, the Wild-cat, the Jersey. There was the Lincoln and the Old Abe and the Sherman and the Yankee. Wells were named after girlfriends and daughters, after geographic features, or simply after their owners. There was, for example, the Ewing well.

Different wells had different personalities. They produced oil and water in different ratios. The oil was of differing quality. Some wells flowed intermittently. One well, it was claimed, only flowed on Sundays. Another well—the Yankee—would stand idle for twenty minutes, then gently cough up spurts of oil and gas for three minutes, then violently belch up oil and gas for some time, and then pause again for another twenty-minute nap. The violent coughing spells could be heard at a distance of six hundred feet.

One out of ten wells struck oil. Nine out of ten wells struck dust.

Of the wells that struck oil, the better producers were easily identified by the rows of blackened tanks, invariably made of wood, sometimes but not always roofed over. In general, the tanks sat close to the creek. In general, they leaked. The crude, black when seen in the shaded dimness of a tank's innards but greenish in the sunshine, flowed in tributary rivulets to Oil Creek. The color of the surface of Oil Creek, where the crude spread into a sheen, was described as "exceedingly delicate and beautiful."

The smell was overpowering.

It was not uncommon for wells to be separated by less than seventy-five feet. I stroll through what was once the McClintock Farm. The first well here was drilled without a steam engine, in 1860. Instead of a steam engine and a rocking beam, the men used a spring pole, a downed tree laid across a fulcrum and rigged in a manner such that two men, by stepping on and off a rope loop, could cause the pole to bob up and down and in so doing cause their drill bit to chisel into the earth. Toward the end of the year, they struck oil. They did not have sufficient barrels on hand to capture the flow and for a time their wealth ran freely downhill and then downstream.

They acquired barrels. They acquired steam engines. They drilled more wells. They rigged an eccentric drive—a single steam engine running pumps at several wells, using steel connecting rods that ran like cables through pulleys in a rusty network, the whole thing moving back and forth, squeaking steel rubbing against steel.

Now some of the steel remains, unmoving and silent. In places, connecting rods still run through pulleys hung on frames in the forest. Elsewhere I find them on the ground, buried under decaying leaves. An engine house still stands, padlocked and silent. I cannot see into the darkness through its windows.

Had I been here in time—150 years ago—I would have heard the steady movement of steam pistons, the squeaking of steel rods, the hissing of oil, the voices of men, the pounding of percussion drilling. I would have smelled oil. My boots would be covered in mud. If I could find lodging, it would have been a shared bed in a hastily built hotel, its parlor floor an inch deep in oily mud and tobacco juice. I would have seen men building barrels. I would have seen and heard teamsters swearing as they drove horses upstream, the horses harnessed to empty scows returning for more oil.

The teamsters were not known for their love of horses. If I had stayed through the winter, I could have watched the men whipping their horses forward, pushing them upstream against the slush and ice that rubbed the hair from the horse's legs and abraded their skin, leaving it raw and exposed to the mud and oil of Oil Creek. At times the creek was a highway clogged with caravans of dying horses.

In 1865, William Wright, a journalist, visited Oil Creek. He was distressed by the treatment of horses: "One's first impulse is to curse the day petroleum was first discovered, and to knock down the barbarians by whom the task of applying the lash has been voluntarily accepted." And this: "The only drawback to the whole is the cruel treatment of the horse, whose lot has been made worse by the great discovery, though its benefits have been felt by whales disporting themselves in the Arctic seas."

Wright's goal, though, was not to protect horses. His goal was to make sense of the oil boom, to understand the region known as Petrolia, as the region that surrounded Oil Creek was by then called. Like others, he was impressed by the quantity of mud that he encountered. He was also impressed by the works of man. "Arrived at the journey's end," he wrote, "I found a discordant, contradictory mass of facts and figures on my memorandum book; and came to the conclusion that, whatever I knew the first day, I knew much less the second, and nothing at all the third." This was a mere six years after Drake's well struck oil.

Wright encountered more than one driller self-educated in geology. Most of them believed the oil came from coal. Oil was coal drippings that found their way to underground reservoirs. But there were other theories. One man, "fussy and seedy-looking," Wright wrote, "avers that the country is of volcanic origin, and is ready to point out certain rents in the hill-tops, through which Vulcan and his helpers found passages for the smoke and cinders of their forges."

No one suggested that the oil might have come from the decay of ancient algae and zooplankton. Which, in fact, it did.

Wright wrote of the processing of crude: "Oil refineries, belching forth clouds of black smoke, or (as is quite common) lying idle, form one of the features in the landscape." And this: "They are for the most part small establishments, each with a capacity not exceeding three hundred barrels per week."

Refining methods were primitive. They involved heating the oil. They involved what Professor Silliman, in his report to Drake's investors, called "fractional distillation." At 300 degrees, about 8 percent of the crude boiled off, turned to vapor. At 360 degrees, 30 percent of the crude was gone, vaporized. At 600 degrees, half the crude was gone. All these vapors could be recaptured. The vapors could be sent into a tube and cooled, yielding purified products like kerosene and gasoline.

At this time, kerosene was the valued product. Kerosene replaced whale oil in lamps. Gasoline, too explosive, too flammable, too deadly, was of no value. It was allowed to evaporate, or it was cast off on the ground, or it was sent into the creek.

I find my way back to Drake's Well and into the museum gift shop. I talk to a woman there about the journalist William Wright and his visit to Titusville, and I buy Wright's book. She asks if I know that Mark Twain passed through Titusville. I had seen, I tell her, a commemorative plaque in town describing the visit, which occurred in 1869, ten years after Drake's success and four years after Wright's visit. The plaque does not say so, but while speaking Twain was heckled by a Titusville drunkard.

I suggest to her that another writer, Charles Dickens, could have crossed paths with Drake himself. Drake lived for a time at a hotel used by Dickens in New Haven, Connecticut, but that was almost ten years after Dickens had passed through. When Dickens passed through, though, he traveled by steam train, and at that time Drake was a conductor. Whether they met is not known.

I buy a souvenir, a vial of crude oil commemorating the 150th anniversary of the Drake well. The Drake well itself no longer produces. The oil in this vial came from downstream, from a

well on the McClintock Farm, reputed to be the longest producing oil well in the world, with a lifespan of fifteen decades and
counting. It is the same oil that is now recirculated at the Drake
well as a demonstration for visitors.

For the McClintocks, oil brought sudden wealth. Among other
things, it financed an adopted son, John Washington Steele,
known also as Coal Oil Johnny. Coal Oil Johnny later described
a boomtown that grew up at the confluence of Oil Creek and a
tiny tributary known as Cherry Tree Run. The town was called
Petroleum Centre. "For pure, unadulterated wickedness," Coal
Oil Johnny wrote, "it eclipsed any town. For open, flaunted vice
and sin, it laid over any other on the map." In Petroleum Centre,
it was possible to stay drunk indefinitely. It was possible to play
cards all night. It was possible to buy the favors of numerous
women. Until later in life, when his money ran out and he gave
up alcohol, Coal Oil Johnny did not necessarily find Petroleum
Centre objectionable.

For those who ran short on funds in places like Petroleum
Centre, sandbagging was a viable means to an end. Sandbagging
involved filling a small burlap bag with sand, swinging it into the
head of a passing man, and taking the wallet from the stunned
man's vest.

The oil that kept the town afloat was harvested quickly. By
1873, Petroleum Centre was abandoned. It disappeared. Its
disappearance coincided with the disappearance, through prof-
ligate waste, of Coal Oil Johnny's wealth.

"If someone were to ask me," Coal Oil Johnny later wrote, "to
pen a sentiment for the benefit of young men who have to face
the temptations of the world, I do not know of anything better
to say to them than 'Tell the boys to drink water.'"

✳

On my table: a teaspoon, my souvenir glass vial of crude oil, matches, and a copy of the text from a flier that accompanied pint bottles of Kier's Petroleum sometime around 1850. This is the same Kier's Petroleum that inspired Drake to drill.

"Kier's Petroleum," the flier says, "celebrated for its wonderful curative powers. A natural remedy! Procured from a well in Allegheny County, Pennsylvania, four hundred feet below the earth's surface."

But Kier did not want to stretch the truth about the medicinal value of rock oil. "That it will cure every disease to which we are liable," his flier says, "we do not pretend; but that it will cure a great many diseases hitherto incurable is a fact which is proven by the evidence in its favor."

Four decades later, when rock oil was an important fuel for lamps and well-known in households throughout the United States, from an 1892 newspaper article: "The 'rock oil' which sold in bottles for medicine was simply the crude oil of today, though there is no question that that found in the Kier well was of the very best. I have taken many a dose of it inwardly." And this: "The petroleum is popularly taken in doses of a teaspoon before each meal, and, after the first day, any nausea, which it may excite in some persons, disappears."

Medicines based on crude oil remain available today. There is Vaseline as a salve. There is T-gel for dandruff and cradle cap. But raw crude oil is no longer prescribed. The raw product, in Kier's time taken orally, is no longer recommended. The medical community unanimously frowns on ingestion.

I open my vial of crude oil and pour its contents into my spoon. It flows freely, like a light grade of motor oil. It forms a quarter-sized pool almost the color of dark chocolate but for its subtle greenish tint. It is opaque, full of energy, the liquid embodiment of sunlight captured in the distant past. The smell fills my room, tarlike, a mixed aroma of fresh asphalt and diesel fuel, primitive, troubling, unwelcome. Its smell is as black as that of burning coal.

I touch a match to the surface of the spoon. Vapors ignite, dancing blue for a moment before burning out.

I move the spoon under my nose and into my mouth. I ingest crude.

The taste is not as ugly as the smell. It is not ugly enough to trigger gagging. It does not excite nausea. But it is not the taste of a healthful balm. It is not quite like anything else I have tasted. I lick the spoon.

My lips and teeth and the roof of my mouth are slick with crude. The aroma finds its way from my mouth to my nostrils. It persists. For an hour afterward, my tongue feels the slickness on the roof of my mouth. Through the night the odor of crude oil reappears, its vapors finding their way from my stomach to my nose.

Mark Twain, in 1905, had a few words of his own for a man like Kier, an advertiser and peddler of medicinal oil. From Twain: "The person who wrote the advertisements is without doubt the most ignorant person now alive on the planet; also without doubt he is an idiot, an idiot of the 33rd degree, and scion of an ancestral procession of idiots stretching back to the Missing Link."

※

Overall it has been a dry spring, and Oil Creek itself will not float a canoe. But nearby French Creek, another tributary of the Allegheny, will. I rent a canoe and cast off into the light current of French Creek.

Long before Drake, the Cornplanter band of the Seneca tribe paddled canoes along these rivers. They occasionally went as far as Pittsburgh to trade furs and other specialties. Among those specialties were hollow logs fashioned into jars and gourds and calabashes containing oil.

By the time of William Wright's visit, French Creek was home to the Henrietta Well, yielding three barrels a day. It was home to the One Well, producing three barrels a day. It was home to the Niedler Well, which flowed on its own accord, without pumping, to yield twelve barrels a day for eight weeks before subsiding. It was home to the imaginatively named Number One Well, Number Two Well, Number Three Well, and Number Four Well. Numbers One and Three were ruined in a flood. Number Two produced only a trickle. Number Four was being deepened.

In addition to wells, French Creek had a refinery. The refinery had two stills. In those two stills, fifty barrels a day were heated, boiling away the unwanted lighter ends—the natural gas liquids, the explosive gasoline, the diesel. The process left behind kerosene and various residues, including thick tar. Regarding gasoline, from the Titusville *Morning Herald* on September 1, 1871: "It is chiefly to rid the petroleum of this dangerous constituent element that refining becomes essential."

Now French Creek carries me over beds of bright green aquatic grass and black freshwater mussels. The banks are blanketed in deciduous forest, here and there interrupted by river homes, mostly small weekenders in need of repair, mostly with canoes of their own sitting in backyards and leaning against sheds. A kingfisher flies over my canoe, and then another. I maneuver around rocks, and a bald eagle steps off its perch high in a tree to glide downstream. Later, when I drag the canoe across shallow gravel, I see another eagle, a two-year-old, its head not yet white.

I see no derricks. If any pumps remain, they are not visible from my canoe. But occasionally I peer through the creek's brown water to see pipes lying on the streambed, pipes of the sort that the oilmen called tubing.

Tubing ran down into active oil wells. It was through the tubing that the oil came to the surface. Over time, tubing corroded or clogged with thick paraffin waxes and had to be replaced. And so some of the old tubing, no longer useful, found its way into the creek.

During the boom, 250 scows could take advantage of a single freshet to float them and propel them down Oil Creek and French Creek to the Allegheny. The voyage from the oil wells to Pittsburgh required three or four days. Occasionally a scow caught fire. A burning scow, tied to a riverbank, could not be allowed to ignite the storage tanks or the wells themselves. A burning scow would be cut loose. Adrift, it would encounter other scows, igniting their contents. As the barrels burned, oil drained into the water, where it floated and continued to burn. In such a fire, the bridge at the town of Franklin burned. The town of Rouseville lost twenty-seven citizens to flames that drifted in from upstream.

I paddle down French Creek and into the Allegheny River and on to Franklin. My canoe carries no petroleum, if one ignores its hull, which itself is petroleum, converted to plastic and molded in a shape that a Cornplanter Indian would immediately recognize.

Farther downstream and in the harbors of the East Coast, it was a small leap from scows with barrels to ships with barrels. In the same year that William Wright visited Petrolia, the sailing brig *Elizabeth Watts* took a shipment of crude oil across the ocean, departing Philadelphia on November 19, 1861, and arriving in London forty-five days later. The square-rigged ship carried her product in 1,329 wooden barrels—901 holding crude oil straight from the well, the remainder holding kerosene. The rich smell of petroleum and the fear of a fire at sea necessitated creative recruiting, the picking up of drunken sailors who would not awaken to their decision until they were on board, shanghaied.

A year later, Philadelphia had exported 239,000 barrels, all of it in wooden barrels originally designed to move whiskey, each holding about forty gallons.

Shippers experimented. They used purpose-built barrels. They fabricated upright iron tanks and tin-lined wooden boxes in cargo holds. The tanker evolved. Seventeen years after the *Elizabeth Watts* sailed from Philadelphia, in the new oilfields that had sprung up on the shores of the Caspian Sea in Azerbaijan, Ludvig and Robert Nobel, brothers of the inventor of dynamite, came up with a tanker design that relied on two iron reservoirs connected by pipes. Their ship is often considered a breakthrough in tanker design, a step toward vessels that carry the cargo within their hull, turning the ships themselves into massive barrels of crude. And their ship was steam powered, a vessel reliant on heat and fire for propulsion loaded with a highly flammable fuel.

The Nobel brothers, somewhat famously, and in the footsteps of Benjamin Franklin, chose not to patent their design.

The name of the first Nobel tanker: the *Zoroaster*.

The fate of the *Zoroaster* is unknown, but the fate of her sister ship, the *Nordenskjöld,* is well documented. In 1881 the *Nordenskjöld* was taking on cargo. A loading hose leaked, sending its fluid contents into the engine compartment. There the fluid encountered kerosene lamps and ignited, killing half the crew.

※

Pipelines, it was realized, could replace teamsters. The cruel treatment of horses and the high price of moving barrels in horse-drawn wagons could be put to rest. The first oil pipeline was built near Titusville in 1863, two years after the *Elizabeth*

Watts sailed. It was a cast-iron line with a two-inch diameter and a length of two miles, joining an Oil Creek well to the Humboldt Refinery in Plumer. Pumps pushed the oil through the pipe. The pumps proved inadequate. A second line was built, also of cast iron, but with a three-inch diameter and a length of three miles. The pumps forced the oil through in sudden spurts. The spurts caused vibrations. Pipeline joints leaked. The line failed.

Teamsters continued to apply the lash.

Later pipelines proved more successful. A two-inch line took refined product two miles to the Allegheny River, where it was loaded onto barges and sent downstream. Another two-inch line carried oil for five miles. A six-inch line, seven miles in length, carried more than seven thousand barrels a day by 1865.

The lash of the teamsters fell with less confidence.

By 1866, the teamsters were losing business to pipelines. The teamsters applied their whips, in the form of arson, to the property of pipeline owners. It started with the burning of two tanks, but the fire was contained. The next day a mob of teamsters gathered. The *Titusville Herald* of April 21, 1866, reported that the mob of teamsters numbered close to one hundred men. They were armed. They moved in.

The owner of the tanks had hired watchmen. The teamsters, according to the *Titusville Herald*, discharged a weapon or weapons. The watchmen fired back.

In the midst of the gunplay, a teamster ignited a storage tank. Another tank caught fire, and another. Five burned. As they burned, they leaked oil. The oil, running across the ground, carried the fire to a railroad platform, which itself burned. Railroad ties caught fire. Four cars rigged to carry oil burned. Four hundred and fifty barrels, awaiting loading, burned. Sixty feet of steel track warped in the heat.

The tanks were replaced. The rails were repaired. The *Titusville Herald* reported that the operation was up and running again within twenty-four hours. Armed patrols protected the pipelines and the tanks.

Over the objections of the teamsters, pipelines proliferated. Pipelines delivered energy at the lowest possible cost. The profit motive sent pipelines from wells to refineries to markets. In 1919, fifty years after the teamster riots in Titusville, the *National Geographic* published an article about the wonders of the oil age. The article did not mention teamsters. It did mention pipelines, claiming that their cumulative length in the United States could circle the equator and continue for another five thousand miles.

※

In 1864, two years before the teamster riots, the U.S. Navy rejected the use of oil as fuel for ships. From a navy report: "The further development of this important substance as a fuel has been prevented by the discovery that when exposed to the air of a confined space at summer temperature, it gives off, even through the open bunghole of a barrel, a gas which, when mixed with the atmospheric air, becomes explosive and detonates with the force of a gunpowder."

Three years later, the navy again considered oil as fuel. From another navy report: "Other things equal, the heat generated by the combustion of one pound of crude petroleum vaporizes 52 percent more water than that generated by the combustion of one pound of the combustible portion of anthracite." Burning oil also required less manpower and freed shovelers of coal for other work.

"The petroleum fire starts into full activity instantaneously," the report stated, "and is as instantaneously extinguished, while the coal fire requires about an hour to attain steady action and as long to burn out." But oil, on a weight-for-weight basis, cost four times as much as coal. And it tended to evaporate before it could be used. The report concluded: "Convenience is against it, comfort is against it, health is against it, economy is against it, and safety is against it." So there was no place in the navy for oil.

But in the same year, 1867, the *New York Times* ran three stories describing navy trials. A navy tug named the *Palos* could reach eight knots under coal power. She was retrofitted with two iron tanks full of petroleum. Pipes fed the petroleum to a boiler. Her top speed exceeded fourteen knots. She burned two barrels per hour.

Speed in navy ships wins battles. Speeches were made welcoming the coming of oil to naval warfare.

Over time, oil replaced coal. In 1920, from the secretary of the navy: "No coal burner can fight an oil burner on anything like equal terms." And even more adamantly: "Oil is the very life of sea fighting."

✳

By 1867, oil was used experimentally in locomotives. Also in 1867, petroleum replaced coal in a fire engine in Boston. When the oil-fired steam-powered fire engine was called into service, it raised a hundred pounds of steam within minutes and averaged ninety pounds of steam for the six hours that it was in action, fighting a fire. It worked side by side with coal-fired steam-powered fire engines. A reporter describing the scene commented on smoke coming from the fire engines: "While the streets leading to the fire were choked up with smoke thrown off by other steamers, scarcely any smoke came from No. 3, using the new fuel."

The engines burned oil, but they burned it in fireboxes that heated water. Steam pushed pistons and turned cranks. They were external combustion engines, and they burned crude oil itself, or kerosene.

Refiners continued to throw away a useless and dangerous by-product called gasoline.

An internal combustion engine burns fuel inside its cylinders. The burning of its fuel, the rapid expansion that comes from sudden ignition, pushes pistons outward, converting heat to motion. Water does not act as an intermediary.

One of the first internal combustion engines was designed by the Dutchman Christiaan Huygens in 1678. The Huygens engine was designed to lift water. Its fuel: gunpowder.

By the 1800s, internal combustion engines were exciting inventors. They burned hydrogen and coal gas and kerosene. In 1876 the German Nicolaus Otto used a four-stroke internal combustion engine to power a two-wheel vehicle, a motorcycle. In 1885 another German, Karl Benz, added a wheel and built the first three-wheel internal-combustion-driven car. Also in 1885, yet another German and an associate of Otto's, Gottlieb Daimler, improved the engine. In 1886 Daimler attached the engine to a stagecoach, creating a gasoline-powered motorcar. It had to compete for road space with horse-drawn carriages, but also with steam-powered cars and electrical cars.

In 1893 a gasoline-powered car showed up in America. Its first drive on a public road was on September 21, in Massachusetts. Three years later, Henry Ford sold his first car. By 1919 Americans drove six million cars. Gasoline was no longer thrown away. It had become something of a necessity. It was sold at every crossroads.

From the author of *National Geographic*'s 1919 article on the wonders of the oil age: "In those earlier days the oil refiner put as much gasoline in his kerosene product as the traffic would allow; today the automobilist complains that his gasoline contains too much kerosene."

National Geographic's author worried about fuel shortages. "Where will my children and children's children get the oil that they may need in ever increasing amounts?" he asked. He speculated about alcohol as a fuel. The alcohol could be made from crops, but it would tie up cropland. He pointed out the importance of coal, and its limitations. "Are there no practical substitutes or other adequate sources?" he wrote. "The obvious answer is in terms of present prices; the real answer is in terms of cost in man power."

I fly home to Alaska, propelled by jet fuel. Another name for jet fuel: kerosene. My seat, propelled by burning kerosene, releases 1,537 pounds of carbon emissions.

Alaska, the nation's coldest state, generates income by selling heat. That heat takes the form of oil, piped eight hundred miles from something like two thousand wells north of the Arctic Circle through a single pipeline to Valdez, on Alaska's southeastern coast. From that single pipeline, the liquid heat goes into tankers. The tankers, for the most part, go to California and Washington. There the heat becomes motion—the motion of snarled traffic, of spinning turbines. In exchange, Alaska receives money. Eight of every ten dollars in Alaska come through that pipeline.

Alaskans keep a fraction of their oil. From my home in Anchorage, I drive to Nikiski, the site of an Alaskan refinery. The three-hour drive, propelled by burning gasoline, is responsible for 176 pounds of carbon emissions.

Captain Cook, with his sailing master William Bligh and their crew of about one hundred men, sailed the converted coal carrier HMS *Resolution* past here around May 25, 1778. Going into the inlet, Cook and Bligh would have seen the site of this refinery off their starboard bow. It would have looked much like the rest of the coast, blanketed in black spruce trees.

Two centuries passed. The refinery was built in 1969.

Originally it was a topping plant, not much more complicated than the stills used by Kier and his contemporaries and built mainly to turn Cook Inlet crude oil and Swanson River crude oil into diesel fuel. When flows of Cook Inlet and Swanson River crude slowed, the plant was converted to handle North Slope crude, transported here in tankers from Valdez three hundred thousand barrels at a time, the sea voyage from the end of the pipeline to the refinery's front door requiring less than a day.

I meet a refinery engineer. He exudes enthusiasm for this refinery. It is more than just a collection of pipes and pumps and tanks and valves. It has a personality. It has character. It changes over time, in some cases because of the refinery engineer's ideas, because of his initiatives. If it becomes upset because someone sends it a bad batch of crude, he takes it personally.

Others before him have taken their relationship with the refining process even further. There was, for example, Jesse Dubbs of western Pennsylvania, who worked during the time when gasoline went from a waste product to a valuable commodity. Jesse named his son Carbon. The young Carbon Dubbs, like his father, was enthralled by the refining process. He gave himself a middle name, Petroleum. Mr. Carbon Petroleum Dubbs had two daughters. He named them Methyl and Ethyl.

Here in Nikiski, the crude moves first into a vertical metal distilling column. The crude, like all crudes, is a mixture of petroleum compounds. It is a mixture of tar and diesel and gasoline and light oil and heavy oil and natural gas. The heavier stuff—the tar and the heavy oils, the stuff that becomes asphalt and Vaseline—has longer molecules, longer chains of carbon and hydrogen. The longer the chain, the higher the boiling point. A very long hydrocarbon molecule will become a vapor at a high temperature and return to a liquid state when that temperature drops. A shorter hydrocarbon molecule will become a vapor at a lower temperature and remain a vapor at lower temperatures.

Dark crude oil enters the column. Flames from natural gas heat the bottom of the column to 740 degrees. The crude boils, and its vapors move upward and cool. As the vapors cool, they return to a liquid state. Not too far up the column, light oils return to a liquid state and are captured on trays. The captured liquid flows out of the column. A little higher in the column, kerosene vapors return to a liquid state and flow out of the column. Closer to the top, gasoline vapors return to a liquid state and flow out of the column. All the way at the top, butane and methane and ethane leave the column.

We drive through the refinery. We look at columns. We look at pipes. We look at valves. We do not see oil. All the oil is inside the columns, the pipes, the valves.

So far nothing we have seen here is different, in principle, from what we might have seen in Kier's refinery. Kier heated oil to produce distillate, and that is all that happens in the column. There are nuances—temperature and pressure settings, boiling the oil in a vacuum—but the basic principle is the same. So far, it is physics, changing from liquids to vapors and back to liquids, with no chemistry involved. It is physics that separates the petroleum compounds.

We move on to another column, a kind of column called a cracker. It looks like a distilling column, but it is not. Here long molecules are cracked into shorter molecules. The process uses heat. Heat crude oil to a temperature of about a thousand degrees, and the large molecules break up. They become smaller molecules. This is not a matter of separating one kind of molecule from another. It is not mere distillation. Tar cracks into light oil, kerosene, gasoline, and butane. And now, with smaller molecules, the volume increases. The forty-two-gallon barrel becomes forty-three gallons and forty-four gallons and forty-five gallons.

To make cracking profitable, it cannot be done with heat alone. Heat alone would change the crude into lighter products, but there would be no control. There would be no efficiency. The cracker we look at is a hydrocracker. At 1,800 pounds per square inch—something like the pressure in a reasonably full scuba tank—certain products from low in the distilling column are mixed with hydrogen. The mix is heated to something like eight hundred degrees. Long molecules break apart. Jet fuel and gasoline are born.

In Nikiski the desired end product is mainly aviation fuel. Alaskans burn more aviation fuel than gasoline, so the refinery engineer pushes his towers and pipes and flames to yield more aviation fuel than gasoline. Gasoline, for Alaskans, remains a by-product. In terms of value, in comparison to aviation fuel, gasoline is almost a waste product.

There is more. Contaminants must be removed. Water needs treatment. Pipes must be warmed. Runaway reactions must be avoided. There is the making of hydrogen by heating natural gas and water to 1,680 degrees in the presence of a catalyst. There is the combining of small molecules to make larger molecules— the creation of gasoline from natural gas. Molecules must be reshaped. The plant was built at a time when natural gas was cheap, and now, when Alaskans are running short on accessible natural gas, there are tricks to conserving the stuff, to saving it. Efficiencies must be engineered in and inefficiencies must be engineered out.

Back in the engineer's office, we talk about the process. He writes on a whiteboard, assuming that I know more about chemistry than I do.

The amazing thing to me, I tell him, is that refineries do not explode with great regularity. "What kind of a lunatic," I ask him, "heats thousands of gallons of diesel, gasoline, butane, propane, and kerosene to hundreds of degrees?"

"Refining," he says, "is more complicated than rocket science. It is harder than nuclear science."

On the wall near his desk hangs a photograph of a nuclear submarine. On this submarine, he once served as a reactor engineer.

✳

Lighter fluid, a mixture of petroleum distillates, is sometimes used as an accelerant by firewalkers. The wood is stacked in a crisscross pattern that allows airflow. The lighter fluid is sprayed across the top of the wood and halfway down the sides. A match or a lighter or a flare might be used to ignite the fluid and the wood. The top of the woodpile burns quickly, creating hot coals and embers. The bottom of the pile, without the accelerant, burns slower, leaving a lasting flame. Fire tenders move the hot coals from the top of the fire onto the ground. They create a sidewalk next to the flames, a promenade of red-hot embers, a thousand-degree footpath.

From Proverbs 6:28: "Can one go upon hot coals, and his feet not be burned?"

For centuries before Tolly Burkan launched the American firewalking movement, feet were not being burned in Argentina, in Australia, in Brazil, in Bulgaria, in Burma and China, in Egypt and Greece and Spain, in Malaysia and Singapore and the Philippines and Thailand, in South Africa. Since Tolly Burkan, hundreds of thousands of people, maybe millions, have walked on hot coals with feet not burned.

Burkan himself walked across hot coals for the first time in 1977, the same year that *Scientific American* published an article explaining, in the author's view, exactly how firewalking worked. By the early 1980s, Burkan talked of the Global Firewalking Movement. It was not a trend. It was not a fad. It was a movement. With the businessman Charles Horton, Burkan launched the Firewalking Institute of Research and Education. In the 1990s, corporations noticed firewalking. American Express and Microsoft and Met-Life set up firewalking seminars for employees.

KFC in Australia discovered firewalking. Reportedly, 180 employees of the chicken-cooking franchise took part in a motivational firewalking exercise. Eleven ambulances responded. Thirty firewalkers were injured. But for the most part, firewalkers walked without injury. They walked across motivational and spiritual flames.

The firewalking movement, to some degree, broke into two schools, one focused on instilling confidence, on motivating walkers from the corporate world, and the other focused on spiritual growth, on personal fulfillment. Both schools burned cedar.

In the background, naysayers proliferated. Scientists dismissed the mystical component. Your mind, they said, has little to do with safely walking across coals. It is all about physics.

From an article by Emily Edwards in 1998, published by New York University and the Massachusetts Institute of Technology: "An awkward craving for transformation produces an exploitable market for spiritual experiences outside orthodox religious establishments, an 'exotification' and appropriation of sacred knowledge." And this: "The mechanism that helps white, middle-class Americans breach logical convention and participate in firewalking and related rituals is a form of mutual pretense, a social ritual engaged to shield people from bleak realities and graceless moments."

I believe in the science, but I want few things more than to be shielded from bleak realities and graceless moments. I want to be motivated. I like feeling empowered. With oil and coal and peat behind me, along with forest fires and cooking and deserts, I long to feel burning coals crunching under the soles of my feet, and I want to feel the hot lick of fire between my toes. But first there is the matter of volcanoes, and nuclear weapons, and the sun. It is not yet time to walk on fire.

Chapter 6

STEAMING MOUNTAINS

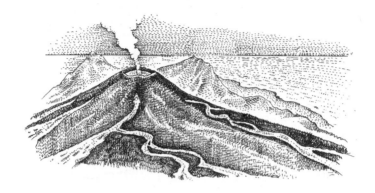

In Hawaii, in a rented bungalow perched high on a volcanic slope, I sweat with fever. The bungalow stands on lava that flowed in 1926, when Mauna Loa was active. The landscape remains wasted. Everywhere, black rock covers the surface. The origin of the rock: pahoehoe flows that moved through as hot tongues of magma, burning ferns and grass and trees as they advanced, their surfaces quickly hardening to a plastic skin and then slowly cooling, steaming for months, becoming this black rock decorated with ripples and odd designs that look like piles of rope or strings of intestine or dark, hard pillows. Elsewhere the flows moved more quickly and tripped over themselves, piling up when they encountered obstacles and fracturing into jagged, sharp, boot-eating edges, 'a'a flows. To fall on 'a'a lava is to rip through denim and lacerate skin. 'A'a flows, even when cool, are best avoided.

I am avoiding them now, too weak with fever to explore, moving in and out of sleep with no sense of time, fighting a virus that appeared out of nowhere. I share the fever and the bed with my companion, her body as feverish as mine. At 4,500 feet our bungalow requires heat, and the heat comes from wood that burns in a small iron woodstove—a Franklin stove. The fuel comes from trees killed by lava and ash and the drought that came here over the last few decades.

Occasionally my companion and I talk or, impatient for recovery, walk around the bungalow, opening curtains to look at the mist, adding a log to the woodstove. The mist is part moisture, part volcanic fumes, gray and vaguely sulfurous. It blankets the hardened lava that flowed from Mauna Loa, along its southwest rift zone, the rift that sent out lava in 1926 and also in 1868, 1887, 1907, 1919, and 1950. Most of the southwest rift flows reached all the way to the sea, falling into the Pacific to steam and pop and add more land, making the Big Island that much bigger.

"The recent lavas of Mauna Loa," wrote Henry Washington in 1923, "include aphyric andesine basalt, chrysophyric oligoclase basalt, and picrite-basalt. The ancient lavas include aphyric labradorite basalt, ophitic olivine basalt, feldspar phyric basalt, and picrite-basalt."

More commonly, to the point of cliché, people wrote of Pele, the Hawaiian goddess of fire and volcanoes.

From Mark Twain in 1866: "We left the lookout house at ten o'clock in a half cooked condition because of the heat from Pele's furnaces."

From Edward Smith in 1885, written in the guest register of the renowned Volcano House hotel: "Pele revealed herself in robes of awful majesty. O Goddess of Hawaiian Lore, enshrouded in the mysteries of eternity, who may know the secrets of thy heart? What scientist may wrest from thy creation or know from whence thou art?"

From George C. Patterson in 1920, also written in the guest register of the Volcano House: "Madame Pele—truly, a most fascinating dame, warm and glowing in disposition, yet fiery in temper, ruddy of cheek and eyes of dancing flames. Quite the most interesting lady I have yet had the fortune to meet."

And from *Time* magazine in 1940: "Pele, goddess of volcanoes, was on a house-hunting expedition when she hovered one day over the Hawaiian archipelago."

❋

Most of the world's volcanoes occur where tectonic plates collide. One plate rides over the other, squeezing magma from between the colliding plates to the surface. Mount St. Helens sits above the collision of the Juan de Fuca and North American plates. Krakatau sits above the collision of the Eurasian and Indo-Australian plates. Vesuvius sits above the collision of the African and Eurasian plates.

Hawaii's volcanoes do not sit above colliding plates, but this in no way lessens their status. They tend to be bigger than the volcanoes of colliding plates. The Big Island's Mauna Kea rises thirty thousand feet from its base on the seafloor to its peak, all lava piled upon itself, flow layered over flow, far taller than the tallest of the colliding-plate volcanoes, taller even than Everest, from base to peak, although its base is underwater. Mauna Loa, also on the Big Island, stands only one hundred feet lower than Mauna Kea, making it, too, taller than Everest.

And Hawaii's volcanoes tend to be hotter than the volcanoes arising from colliding plates. Their heat comes from deeper in the earth.

Somewhere beneath me, beneath this mountain of hardened lava, beneath the earth's crust, convection currents move the very stuff of the planet's inner self upward and downward in spiraling loops. The crust on which we live, seemingly stable, is nothing more than an onion skin, a shell around the convection currents that underlie all the earth's surface. It is convection that has gone on for billions of years in material with a consistency far removed from daily human experience. The currents move molten rock under extreme pressure, a slow-moving fluid, hotter than the hottest ovens, hotter than the hottest smelter, hotter than a blast furnace, but extremely dense, with large quantities of iron, untouchable, something like a very hot and very thick syrup but far more viscous and far less translucent. The roots of the convection lie three thousand miles down, near the earth's solid inner core—rendered solid not because it is somehow magically cooler than the surrounding molten syrup but because of the weight of the thousands of miles of earth above.

The syrup above that solid core—the thick fluid stuff that makes up the entire outer core and mantle—behaves something like the earth's atmosphere. But the heat that powers movement in the atmosphere comes from the sun, while the heat that powers movement in the outer core and the mantle comes from the energy of collisions that occurred at the time of the earth's creation, at a time when space junk collided with space junk to form a cohesive mass that fed on more space junk to become a planet, with each collision adding a bit more heat, each collision like a hammer blow that heats the head of a nail, creating movement in its molecules, movement that is nothing more and nothing less than heat. That heat has been escaping for four billion years, leaking slowly.

The heat comes, too, from radioactive rocks. Radioactive elements are heavy elements. Because of their weight, they settled in the core during the formation and aging of the earth. In the core, these heavy elements break down, their nuclei splitting to form lighter elements, smaller atoms. And in breaking down, they release heat.

Just as air circulates in the atmosphere, so viscous liquid rock circulates in giant spiraling convection cells, and the tectonic plates float on top of these cells, moving at the speed of a growing fingernail, riding the currents created by heat. And within one of those circulating cells, the Hawaii hot spot behaves something like a thunderstorm, sending a column of heat upward, right through the middle of a plate, where it cools and then falls back downward, a local phenomenon, but intense. It is a storm with a diameter of forty-five miles, a storm that stays in place as the earth's crust drifts past. It is a stationary storm that pokes through the moving tectonic plate to create a chain of islands, the newest, the Big Island, to the east, the oldest, Kauai and Niihau, to the northeast, eroded and no longer active. Further along, beyond what is thought of as the Hawaiian Islands, stretching to the west and north, more islands exist, even more eroded than Kauai and Niihau, some now barely above the water, like Midway and Kure Atoll, or underwater altogether, forming a chain that stretches toward the Russian peninsula of Kamchatka, toward the edge of the tectonic plate. The line formed by these eroded mountain islands traces the movement of the tectonic plate above the stylus of the stationary hot spot, the stylus of the stalled storm that originates near the core and sends its heat upward to penetrate and scar the crust.

The hot spot's heat rises from somewhere well above the core but well below the earth's surface. As the molten current reaches upward, it finds the bottom of the lithosphere, the hardened outer part of the mantle just below the crust, the bottom layer of the onion skin. The heat that originated near the core, that sent updrafts of molten rock, melts the lithosphere at a depth of fifty miles. That melted rock collects in a magma chamber beneath the earth's crust. A vast reservoir forms. And that magma chamber flows through a conduit that stretches twenty miles, from the top of the magma chamber to the surface, right through the floating tectonic plate and the thin-skinned crust, seeking a way out, an opening, a vent.

❋

My companion and I drift in and out of fever dreams. When awake at the same time, we drink tea and talk of lava and history and heat. Next to our bed, amid a heap of half-read books, I find my infrared thermometer. I point it at the bare skin of my thigh. "One hundred and two," I tell my companion. Her own skin comes in at 101.

We talk of fever in the islands. The Hawaiians suffered almost as badly as the Alaskans. In 1778, when Captain Cook showed up in his converted collier, something like a half million islanders lived in a highly structured agricultural society. But Cook's collier carried disease instead of coal. By 1854 a census counted seventy-three thousand people, including two thousand foreigners. By 1890 fewer than forty thousand native Hawaiians remained.

The Hawaiians, before Cook, learned about the sacred living force, which they called *mana*. They learned about harmony or righteousness, which they called *pono*. They learned about unity with the universe, which they called *lokahi*. They learned that *mana*, *pono*, and *lokahi* together formed the triad of health.

From Cook's men, they learned about syphilis and gonorrhea. They learned about morbidity, mortality, and infertility. In 1804 Hawaiians learned that cholera or typhoid could kill fifteen thousand people in a single epidemic. In 1839 it was mumps. In 1840, leprosy. In 1845 they began a four-year lesson in influenza. In 1853 and 1854, it was smallpox. In 1896 the imported Asian tiger mosquito began offering classes in dengue fever. In 1899 school was in session for bubonic plague.

As a matter of public health, officials established quarantines. They torched infected buildings in Honolulu. On January 20,

1900, they lost control of the fire on Beretania Street and burned down Honolulu's Chinatown.

The fevers in our rented bungalow suddenly seem irrelevant, our complaints self-indulgent. Dehydrated and shaky, we venture outside, picking our way across the lava landscape of our three-acre compound. Plants grow in scattered patches. Small ferns grow from cracks in the lava, bits of shade that might capture moisture, and there are red-flowered 'ōhi'a lehua trees and shrubs, known for their tolerance of heat and their ability to recolonize after fires. The larger trees do not bear leaves. They are long dead, dried and bleached in the sun, their light gray wood striking against the black lava. It is these trees that fall and become firewood for our stove. Scattered dead tree ferns speak of wetter times, rain forest wet, dripping. I find a tree mold, the cast of a tree formed by hardened lava, like a hollow log but made from rock. I find another and then another, and another, a small thicket of tree molds where hot lava flowed around trees and hardened before the trees themselves ignited and burned away. Each mold stands two or three feet above the surface. I can peer down into their hollow cores, into the darkness well below the surface. It is like looking back in time, to 1926, seeing down to the ground's old surface. It is a reminder that the ground here is not what it seems. The ground on which I stand is neither old nor predictably permanent. A few years before my father was born, this rock emerged from the earth and flowed, heated by the flames of a furnace three thousand miles below.

My companion takes pictures. She complains of the difficulties of photographing black lava, of the lack of contrast, of the camera's confusion with regard to light. I look back in time, peering down into tree molds. I find one in which a new tree has taken root. A young 'ōhi'a lehua tree grows from the death mask of an ancestor.

Back in the bungalow, weakened by our excursion, I work on plans to see fresh lava, liquid earth, rock caught in the throes of fever. I want to run a spoon through fresh lava and feel its viscosity.

I telephone my contact at the Hawaiian Volcano Observatory, a young geologist known for his work with infrared imaging of volcanoes. He has bad news. The volcano is quiet. It is exhibiting a deflation event, meaning that the lava has retreated into the vent, has subsided somewhat in the conduit that connects the surface to the magma chamber below. The bitch Pele is not cooperating.

❋

Papandayan in Indonesia killed just fewer than three thousand people in 1772. Mount Pelée in Martinique took almost thirty thousand people in 1902. El Chichón in Mexico killed two thousand in 1982. These were deaths, for the most part, from hot ash and flows of hot gas mixed with rock, known as pyroclastic flows, or pyroclastic density currents, or simply PDCs. But volcanoes have other ways to kill. There are mudflows: 25,000 dead in Ruiz, Colombia, in 1985. There are tsunamis: 36,000 dead in Krakatoa, Indonesia, in 1883. There is starvation in the aftermath: 9,000 dead in Laki, Iceland, in 1783, and 92,000 dead in Tambora, Indonesia, in 1815.

Pliny the Younger, nephew of Pliny the Elder, wrote about the Vesuvius eruption that destroyed Pompeii in AD 79: "Gross darkness pressed upon our rear and came rolling over the land after us like a torrent." He wrote of pumice and scorched rocks falling onto his ship anchored offshore. "Broad sheets of flame were lighting up many parts of Vesuvius. Their light and brightness were the more vivid for the darkness of the night. To alleviate people's fears my uncle claimed that the flames came from the deserted homes of farmers who had left in a panic with the hearth fires still alight."

Pliny the Elder went ashore south of Pompeii, in part to investigate the eruption and in part on a rescue mission. He headed, his nephew wrote, "straight for the danger zone." Ash was falling, and fires continued to burn on the hillsides. The people were panicked. Pliny the Elder—again from his nephew's account—took a bath and then slept. He was a large man, and his companions heard him breathing as he slept. They woke him when ash accumulated in the streets and in the courtyard outside his room. They strapped pillows to their heads, protection from falling pumice. They moved toward the sea. Pliny the Elder was not in the best of health. Today he would be considered grossly obese. On the beach, he collapsed and died. The volcano, aided by Pliny's poor condition, killed him.

Thousands more died from heat and ash and fumes. Much later, scientists found rocks around Pompeii that had been hot enough to melt lead. The ash cloud billowing up from the ground exceeded 1,500 degrees. In nearby Herculaneum, the heat came so quickly that people had no time to panic. Human remains, preserved as imprints in volcanic ash, are frozen in various relaxed postures.

Vesuvius erupted again in AD 172, in 203, in 222, in 303, and so on, until 1944, with varying degrees of ferocity. For some time now, it has been quiescent. Today three million optimists live in its blast zone.

Two days later, we continue to wait for lava. I talk to the geologist at the Hawaiian Volcano Observatory. Satellite images show activity on the surface, he says. Instruments in the throat of Kilauea, the currently active volcano, show nothing but deflation.

"Don't worry," he tells me. "These deflation events never last long." But Hawaii is dotted with dormant volcanoes. It is easy to say that Kilauea is alive and well, that its deflation will not last, but in fact it could go quiet for a long time. Forever. And we have only ten days left on the island.

Kilauea means "spewing" or "spreading" in Hawaiian, as in spewing ash or spreading lava. The volcano is said to be the most active volcano on the planet, with more or less reliable activity since 1983, and accounts of activity going much further into the past.

Most of the activity is contained within the huge and inaccessible pit of a caldera, magma roaring up, cooling, and then sinking back, sending up nothing but fumes. But some of the activity involves lava easing out of cracks in the side of the volcano, well removed from the crater, below and to the east of its summit, close to the sea. If we are to see live lava, we will see it at the east rift zone, but only after the deflation event ends and the lava reaches a level that allows a bit of it to overflow through the cracks in the mountain's surface, openings from which flows issue forth to create new surfaces, to create new layers on the shield, which will themselves eventually be covered by younger surfaces, one covering another for as long as the mountain remains alive.

Waiting, we hike into the Ka'ū Desert, a place spewed over by Kilauea, but also by Mauna Loa. We move slowly, still fever worn. The route takes us over 280-year-old rock, and then over 420-year-old rock, young earth in most places, but respectably old in this part of Hawaii. Ash and grit from 1790 cover certain areas, and we hope to find 220-year-old footprints, the footprints set down as ash fell from the sky—footprints of men and women who may have met James Cook himself.

The eruption of Kilauea in 1790 may have killed as many as five thousand Hawaiians, or as few as eighty. They were part of an army of warriors, a raiding force crossing the island, or they were workers and their families, or both. Sheldon Dibble's account, written five decades after the eruption, described one party of warriors catching up to another, only to discover that they were all dead. The warriors traveled with their wives and families. "Some were lying down," Dibble wrote, "and others were sitting upright clasping with dying grasp, their wives and children. So much like life they looked, that they at first supposed them merely at rest, and it was not until they had come up to them and handled them, that they could detect their mistake." Oral histories and tourist brochures repeat Dibble's account, but most of the footprints are too small to have been made by warriors.

A roofed display covers a patch of footprints, but the footprints are badly eroded, better described as foot smears. Without the interpretive sign, I would not recognize them as footprints.

We search for more footprints but find none. We wander upslope across black rock. We step over cracks with the now familiar ferns and 'ōhi'a lehua shrubs. In places, our footsteps ring hollow. The ground beneath is an empty shell, a facade of hardened lava over air-filled space where hot rock exposed to air cooled, and hotter rock beneath it continued to flow and drained away.

We find a place where the hardened outer rock has formed a ledge, a shelf, with the dome of rock covering a cave that extends back twenty feet, a giant bubble of black rock. We find pisolites, beads of lava the size of BBs, formed when electrostatic charges pull together bits of ash inside the eruption cloud.

The volcanoes in Hawaii take the shape of warrior's shields, shallow hills rather than sharp cones. After two miles, we climb to the top of Mauna Iki, a lava shield born in 1919. The shield grew for eight months. It is a scale model of the larger volcanoes of Hawaii, Mauna Loa, Mauna Kea, Kohala, Hualalai, and Kilauea. Mauna Loa, visible from here, is the largest shield volcano on earth, lava piled on lava piled on lava. Mauna Iki, under our feet, is small enough to take in at a glance. It is really part of Kilauea, part of its southwest rift zone, the lava of Mauna Iki coming from an offshoot of the Kilauea plumbing system. Near the summit, a pit crater opened where a lava pond drained back into the plumbing system, letting the ground above collapse.

I find a fissure. If we are to see live lava, if the deflation event ends, the lava we see will come from fissures like this. The fissure is two feet wide at the surface, but it narrows as it descends into the darkness of the earth, deeper than I can see. The lava that once flowed from the fissure drained away before it hardened, falling back into the plumbing system to leave this open crack. I want to climb into the fissure, to explore it as I would explore a cave, but when I put my hands on the ground, I abandon the idea. The surface is sharp where tiny bubbles of molten glass had formed and burst. It is sharp enough to tear pants and shirts and the palms of hands.

I crouch next to the fissure and look into its darkness, imagining it full of lava, glowing hot and red and sulfurous, overflowing to build the shield of earth on which we walk, overflowing to create the ground beneath our boots.

Shield volcanoes tend not to explode. Their lava boils over slowly and flows at rates that can be outwalked, even by the infirm. When a shield volcano sends lava flowing toward your house, you will, in all likelihood, have time to remove your furniture.

Stratovolcanoes—volcanoes that form majestic cones—explode. Of the 1,511 volcanoes believed to have erupted in the past ten thousand years, 699 were stratovolcanoes. Papandayan in Indonesia was a stratovolcano, as was Mount Pelée in Martinique, and El Chichón in Mexico, and Krakatoa in Indonesia, and Vesuvius.

Novarupta in Alaska is a stratovolcano that exploded in June 1912. It eviscerated something like three cubic miles of material, projectile vomiting in a manner that could be heard more than seven hundred miles away. By volume of material released, it was the largest eruption of the twentieth century. It happened in the back of beyond, well north of Anchorage, in the middle of nowhere, far from cities, but its ash fell as far away as Seattle.

Closer to Seattle is Mount St. Helens, another stratovolcano. Its eruption in 1980 very suddenly killed fifty-seven people and wiped out hundreds of homes, forty-seven bridges, fifteen miles of railway, and 185 miles of highway.

Mount St. Helens offered warnings before it exploded. The earth swelled. The ground trembled.

Just before the mountain blew up, Stanley Lee, a longtime resident and store owner, sixty-seven years old, had this to say: "It's just a crock cooked up by the federal forestry service or them environmentalists to delay a big development of the Spirit Lake recreation areas."

A photographer was killed, his car half buried in ash, its windows gone. A geology student lived long enough to leave footprints in the ash that marked a looping path out and back to the place where he died, of asphyxiation, with ash in his throat. Sixteen others died in a similar manner. One man died eight miles from the explosion when a rock crashed through the window of his car. Twelve miles from the volcano, three died from burns.

If you want to visit a volcano, if you want to see flowing lava, to feel its heat, it is wise to focus on shield volcanoes.

❋

I call the Firewalking Institute of Research and Education. I reach the instructor. He is planning a training course in Argentina and another in California.

I ask him about walking on lava. "Charles Horton did it," he tells me. Charles Horton is one of the big names in the firewalking movement, a name almost as well known as that of Tolly Burkan himself. Horton, the instructor tells me, was on a helicopter tour above the lava. He convinced the pilot to land so that he could walk on new ground. He took off his shoes and strolled across ground that had only recently lost its incandescent glow.

The instructor clearly envies Horton's lava-walking experience. He hopes and plans to do it himself one day. He dreams of being surrounded by glowing rock, of feeling the heat beneath his feet and in the air around him.

"It would be mind blowing," he says. And I agree.

※

Another day passes without lava on the surface.

In a rented convertible, we drive along the coast to the rainy side of the island, across the lower flank of the shield that is Mauna Loa, the earth's largest active volcano, shorter than its inactive neighbor, Mauna Kea, but more massive, more voluminous, with something like ten thousand cubic miles of rock to its name.

Along certain stretches, the lava of the Mauna Loa shield is old and covered with rain forest, thick stands of dripping tree ferns and an abundance of bright flowers. Along these stretches, it is hard to see Mauna Loa as a volcano. The shield slopes lazily upward to the left and downward to the right, toward the sea, looking no more volcanic than Vermont.

Charles Darwin described a stratovolcano's beautifully formed, smoking cone:

> The ruins of Concepcion is [sic] a most awful spectacle of desolation. There absolutely is not one house standing. I

have thus had the satisfaction in this cruise both of seeing several Volcanoes & feeling their most terrible effects. It is certainly one of the very grandest phenomena to which this globe is subject.

At about the same time, Herman Melville wrote skeptically of a shield volcano in the Marquesas Islands, two thousand miles from Hawaii:

That the land may have been thrown up by a submarine volcano is as possible as anything else. No one can make an affidavit to the contrary, and therefore I will say nothing against the supposition: indeed, were geologists to assert that the whole continent of America had in like manner been formed by the simultaneous explosion of a train of Etnas, laid under the water all the way from the North Pole to the parallel of Cape Horn, I am the last man in the world to contradict them.

In places, though, the route along the Mauna Loa shield changes from rain forest to bare and nearly bare hardened lava, unmistakably volcanic. After we pass through Hilo, turning west to connect with the Saddle Road and gaining altitude, we enter an almost lunar landscape of black rock. We move steadily upward on new pavement, a very good road. In clear patches we see the telescope domes on top of Mauna Kea, staring upward, but we pass into banks of fog and clouds and rain and are forced to put the top up, to seal off our convertible from the outside.

While stratovolcanoes may be more dangerous in terms of sudden explosions, of fast-moving pyroclastic flows, of overwhelming deposits of hot ash, shield volcanoes can still ruin your day. Slow-flowing lava leaves time to move the furniture out of the house, but the house still burns. Lava from Mauna

Loa moved toward Hilo in 1935 and 1942. The lava formed tubes and troughs that conserved its heat, allowing it to flow long distances without hardening. The military was called in to bomb the flows, to break open the walls of tubes and troughs so that the lava would cool and harden or at least flow elsewhere, to someone else's backyard. The lava, ignoring the bombings, stopped on its own. Hilo, in 1935 and 1942, was safe.

❋

Twenty-one miles along the Saddle Road, high on the shoulder of the shield, we turn left onto a single-lane road of cracked concrete and potholes. We continue upward and leave all vegetation behind. We break out of the clouds above the rain. Now there is nothing but black and brown and yellow-tinted pahoehoe flows and scattered ʻaʻa flows, with lava on top of lava, flows from different events lying one on top of another, and on top of all of them this winding, wounded track that discourages speeding without the need for posted limits.

We stop and put the top down.

At the end of the road, we park beneath a sign that says "Observatory Trail." This is not the astronomical observatory of Mauna Kea, which is north of here, on the other mountain, the neighboring volcano. This is the Mauna Loa Observatory, an atmospheric observatory.

But it is not just an atmospheric observatory. Arguably, it is *the* Atmospheric Observatory. It was here that Charles David Keeling kept his instruments. It was here that Keeling's data grew into the Keeling Curve, not a curve at all but an icon, not a curve at all but a warning, not a curve at all but a climbing staircase of carbon dioxide over time, a graphical fact. It is a staircase climbing from 290 parts per million in the eighteenth century to about 310 parts per million when Keeling started his work in 1958 to close to 380 parts per million in 2005 when Keeling died, and nearly 390 parts per million today.

A man shows us around the facility. He has worked here since carbon dioxide levels were below 360 parts per million. That is to say, he has worked here for seventeen years.

We look at a plaque bolted to the side of a building, proclaiming that this is the Keeling Building. The building is more of a glorified shack than an institutional monument. It would be fair to say that the plaque is the Keeling Building's finest feature. The plaque includes an embossed version of the Keeling Curve, from 1958 until 1997. The curve is an inclined, saw-toothed line, headed upward. The sawtooth pattern comes from seasonal changes in carbon dioxide. Levels go up in the winter when plants in the northern hemisphere senesce, and they go down in the summer when plants in the northern hemisphere are active. The overall trend is overwhelmingly upward. The data are so clear that they appear contrived. But they are not. They are measured data.

Above the Keeling Curve, the plaque reads: "Keeling Building: Named in honor of Professor Charles David Keeling, Scripps Institution of Oceanography, who initiated continuous CO_2 measurements at this site in 1958." The plaque is perhaps fifteen inches long by twelve inches tall.

Before Keeling came Jean Baptiste Joseph Fourier and John Tyndall, when carbon dioxide levels hovered around 290 parts per million. But these men were not so much concerned with human-induced climate change as they were with explaining the comings and goings of the glaciers that had once buried so much of what had become civilization.

Also well before Keeling, the Swedish chemist Svante Arrhenius took things one step further, working for months with a pencil and paper on mathematical calculations. Arrhenius had read Fourier's work, and Tyndall's. He began his 1896 paper with a simple statement: "A great deal has been written on the influence of the absorption of the atmosphere upon the climate."

And this: "I should certainly not have undertaken these tedious calculations if an extraordinary interest had not been connected with them."

That extraordinary interest was not future climate change but past climate change. Arrhenius's interest—like that of Tyndall—stemmed from causes of past ice ages and past warm periods. "From geological researches the fact is well established that in Tertiary times there existed a vegetation and an animal life in the temperate and arctic zones that must have been conditioned by a much higher temperature than the present in the same regions. The temperature in the arctic zones appears to have exceeded the present temperature by about 8 or 9 degrees." He computed that a doubling or tripling of carbon dioxide levels would raise temperatures about nine degrees, back to the levels he associated with the Tertiary. "In the Physical Society of Stockholm," he wrote, "there have been occasionally very lively discussions on the probable causes of the Ice Age." He calculated that roughly halving the levels of carbon dioxide would bring a return of the ice ages.

Arrhenius's work faded into a footnote. Other explanations, more plausible explanations, were offered for the ice ages. Then, in 1938, came the English steam engineer Guy Callendar. "Few of those familiar with the natural heat exchanges of the atmosphere," he wrote, "which go into the making of our climates and weather, would be prepared to admit that the activities of man could have any influence upon phenomena of so vast a scale. In the following paper I hope to show that such influence is not only possible, but is actually occurring at the present time." He blamed the influence on fossil fuels: "By fuel combustion, man has added about 150,000 million tons of carbon dioxide to the air during the past half century." He believed that he had data, collected from two hundred weather stations, showing a small increase in temperature.

But Callendar was no poster child for today's climate change activists. He welcomed a warming climate. "In conclusion," he wrote, "it may be said that the combustion of fossil fuel, whether it be peat from the surface or oil from 10,000 feet below, is likely to prove beneficial to mankind in several ways, besides the provision of heat and power. For instance, the above mentioned small increase of mean temperature would be important at the northern margin of cultivation."

The warming would protect us from another ice age. In his words: "The deadly glaciers should be delayed indefinitely."

His work was ignored. People were not ready to believe that humans could change the earth on such a monumental scale. Naysayers said his carbon dioxide measurements could not be trusted. They said his temperature measurements were flawed. The difficulty with Callendar's thesis—one of the difficulties with his belief that humans could alter carbon concentrations in the atmosphere—was the ocean. Naysayers latched on to the world's oceans. The oceans were vast enough to absorb tremendous amounts of carbon dioxide. The oceans would absorb carbon dioxide produced from fuels.

The oceans washed Callendar's work into obscurity. Callendar became a sidebar in textbooks, a footnote.

Enter oceanographer Roger Revelle, twenty years later, in 1957. Revelle ran the Scripps Institution of Oceanography. Scripps then was a mere foreshadowing of Scripps today. It was Revelle who led the growth of Scripps into one of the best-known oceanographic laboratories in the world. Revelle was also an expert on the esoteric chemistry of carbon and calcium compounds in the ocean. With his knowledge, he was one of nearly a hundred scientists and technicians sent to Bikini Atoll to assess the effects of the hydrogen bomb test. Revelle assigned a few of the scientists to look at water chemistry. He was interested, too, in ocean mixing. He was interested in understanding the oceans as a potential receiving basin for radioactive waste. He studied the effects of a nuclear bomb used as a depth charge, a submarine killer, finding, among other things, that the contaminants from the bomb did not move quickly through the water column. Waste put into one layer of water might not readily spread to another. And carbon dioxide absorbed at the surface might not readily spread to the deeper layers. Most of the molecules of carbon dioxide finding their way into the surface waters would actually also find their way right back out, into the atmosphere.

Revelle's work dismissed the naysayers who had dismissed Callendar. Revelle's work reopened the possibility that carbon dioxide from the burning of fossil fuels was accumulating in the atmosphere. And it was Keeling's follow-up, beginning around the same time that Revelle was publishing and talking about his work, that showed what was in fact happening. Keeling's collection of samples and consistent measurements of carbon dioxide at the end of a long, rough road near the top of Mauna Loa showed, beyond doubt, accumulation of carbon dioxide.

Samples from other locations backed up Keeling's work. There were samples from ships far at sea. There were samples from the whaling community of Barrow at the northern tip of Alaska. There were samples from the South Pole. There were samples from American Samoa and California. And in the end there was a plaque bolted to a glorified shed near the top of a volcano and an understanding that something big was going on.

We go inside another building, bigger than the Keeling Building but hardly grandiose, a building that would be on the small side in a suburban light industrial park. Inside, in a hallway, amid a clutter of worn-out government-issue furniture, sits Keeling's black box. It is the box that contains the original instrument he used to measure carbon dioxide levels. It was replaced around the time of Keeling's death, in 2005. It is a shrine.

We walk down the path to our car and drive down the mountain, across lava fields, black and gray and never changing, but never monotonous. In driving, we add 150 pounds of carbon to the atmosphere. Our flights, Alaska to Hawaii and back, in tourist class seats, account for another three thousand pounds.

※

Still no lava. "But tomorrow looks promising," the geologist tells me, "or the day after. The satellites are picking up warm spots." Warm spots mean the lava is close to the surface, but not quite there.

We sit in the sun outside our bungalow, perched on lava. The public relations machinery of the climate change movement makes it easy to think that climate is all about carbon dioxide, but the story is more complex than that. It is also about the sun and the earth.

The sun's brightness changes. There was the Maunder Minimum, from 1645 until 1715, with very low sunspot activity and cooler temperatures here on earth. The astronomer John Eddy wrote about the changes in the sun's temperature in 1976: "The reality of the Maunder Minimum and its implications of basic solar change may be but one more defeat in our long and losing battle to keep the sun perfect, or, if not perfect, constant, and if inconstant regular," he wrote. "Why we think the sun should be any of these when other stars are not is more a question for social than for physical science."

He wrote, too, of the similar Spörer Minimum, from 1460 until 1550, and the Dalton Minimum, from 1790 until 1830. Greater sunspot activity, it seems, consistently means warmer climates. Sunspot quietude means cooler climates.

And there are the Milankovitch Cycles described by Milutin Milankovitch, cycles based on the combined timing of changes in the earth's orbit, the angle of the earth's tilted axis, and the direction of the axis during the earth's closest approach to the sun. The earth's orbit flattens and broadens regularly, every hundred thousand years, changing the difference between summer and winter sunlight by something like 20 percent. The tilt of the earth's axis changes, making a sharper angle toward the sun, regularly, every forty-one thousand years, adding to the difference in sunlight received at different parts of the earth in summer and winter. And the summer solstice—the time when the axis angles most directly toward the sun—coincides with the point in time when the earth is at its closest distance to the sun every twenty-three thousand years, adding even more to the differences between summer and winter.

None of these changes affect the total amount of sunlight received by the earth each year, but they change the seasonality of sunlight and heat, and the parts of the earth receiving the most heat, which in the end changes climates. During World War 1, Milankovitch, interned by the Austro-Hungarian army, began the tedious mathematics needed to understand these cycles, and two years after the war he published his *Mathematical Theory of Thermal Phenomena Caused by Solar Radiation*. Two decades passed, and just as a new war was breaking out, he published *Canon of Insolation of the Earth and Its Application to the Problem of the Ice Ages*. Milankovitch Cycles, he believed, had much to say about changing climate.

Over longer periods, climate change is even about the position of the continents and the locations of mountains. The continents guide water currents, both warm and cold. The mountains guide air currents. And they come and go. The earth of fifty million years ago was not the earth of today. The sun of 1700 was not today's sun. The geometry of the solar system experienced by *Homo erectus* was not the geometry of the solar system experienced by *Homo sapiens*.

When Jean Baptiste Joseph Fourier wrote of a greenhouse effect in 1824 and 1827, he did not know of the complexities that would surface. In 1861, when John Tyndall reviewed climate change in ninety-five words, he did not understand all the factors in play. Neither Fourier nor Tyndall dwelled on sunspots and planetary geometry. They dwelled instead on the atmosphere as a blanket. Now, nearly two centuries later, ten thousand details remain to be worked out. But this much is clear: we have 40 percent more carbon dioxide than we did during the Maunder Minimum, 29 percent more carbon dioxide than when Milutin Milankovitch was born, and 17 percent more than when John Eddy wrote of the defeat of a perfect sun. And the world is getting warmer.

For now, astronomical cycles work on our side, counteracting some of the warmth trapped by a thickening blanket of carbon dioxide, but despite that, temperatures are rising.

My companion and I emit another eighty pounds of carbon driving our convertible to Kilauea's caldera, the huge sunken kettle near the mountain's summit. The caldera is the centerpiece of Hawaii Volcanoes National Park.

Edward Abbey complained of Smokey Bear but thought of the National Park Service as a good employer. "I love my job," he wrote in his famous book *Desert Solitaire*, published in 1968. This did not mean that he loved his employer's decisions. He loved his job because nature surrounded him and the limited workload gave him time to think. "What little thinking I do is my own," he wrote, "and I do it on government time."

One of the things he thought about was roads cut into the wilderness to allow easy access, like the one on which we drive.

The roads attract tourists in droves but repel the likes of Abbey, and for that matter me, men who prefer to work for their scenery. In this case, though, the attraction outweighs the repulsion. I remain slightly feverish, which rules out real hiking, and there is a volcano at stake. And not just any volcano but the world's most accessible volcano, Kilauea. I use the road.

Kilauea's caldera, although not always as easily reached as it is today, has attracted tourists for some time. When Mark Twain wrote of the goddess Pele and the lookout house, he was writing about Kilauea. "As we 'raised' the summit of the mountain and began to canter along the edge of the crater, I heard Brown exclaim, 'There's smoke, by George!' (poor infant—as if it were the most surprising thing in the world to see smoke issuing from a volcano)." Twain goes on to claim that he was disappointed by what he saw. "Only a considerable hole in the ground," he wrote, "it is a large cellar—nothing more—and precious little fire in it."

We park our convertible and approach via a short walk through a forest of tree ferns and ʻōhiʻa lehua trees. Thick steam rises from a crack next to the trail. "Look," I say, "steam!"

We round a corner, take three steps through thick brush, and the crater appears abruptly.

Kilauea's caldera is nothing short of remarkable, even when quiescent, as it was when Twain visited. It is roughly circular and two to three miles wide, somewhat oblong rather than circular, but thirteen times larger than the Nuclear Testing Ground's Sedan Crater and six times larger than Ubehebe Crater of Death Valley. And it is young. It may have formed within the past few hundred years when magma drained from beneath the summit of the shield and the whole thing collapsed, like limestone collapsing into a cavern to form a sinkhole but on a larger scale, and in black basalt instead of white limestone.

For many years, visitors came here to look down into a lava

lake or lakes. Glowing magma flooded the floor of the crater. "My first visit was on the twenty-third of May 1864," wrote George Clark in July 1867. "At that time there was but one lake of any note. On my last visit the first thing that attracted my special notice was the large north lake entirely new to me as were also several other lakes."

"Crater filled with several 'floating' islands all surrounded by spouting fountains of yellow lava," wrote F. F. Woodford in 1921.

Others described watching a hard crust of cooled lava break apart. "Another and another piece would break off, rush forward, and disappear the same as the first, until the entire surface would be broken up and submerged," wrote Charles Marlette in 1865. "After this violent action had continued from fifteen to thirty minutes and the old crust had all or nearly all become melted or submerged the lake would gradually become more quiet."

Although often called a crater, the pit at the top of Kilauea is better called a caldera, a magma chamber that has collapsed. The caldera floor, once wide awake with pools of molten lava, dozed off starting in the 1920s. No explosive eruptions were observed for the two and a half decades starting in 1982. Kilauea's caldera became a pit of steaming black rock, or hardened magma, dramatic in a static sense, but not the seething witch's cauldron that it had been. Visitors could drive around a Park Service ring road that circled the crater. They could park and walk to an observation platform and look down into the pit. They could toss pennies over the railing and watch them disappear into the distance.

Then one night in March 2008, a steam vent exploded. It scattered broken rocks over seventy-five acres. It threw a boulder onto a fence near the viewing platform. Its ash covered part of the ring road. It opened a crater in the bottom of the caldera.

And this crater, called Halemaʻumaʻu, the home of the goddess Pele, continues to send up steam and sulfurous smells. In the bottom of the crater, lava boils to the surface, cools, and sinks back down. When the steam clears, or with infrared scopes capable of peering through the steam, the lava flows can be seen moving like water in a violently boiling kettle, forming a wave that rises from below on one edge and sinks on the other, circulating molten earth in a giant convection current.

As a general matter, Kilauea is either deflating or inflating. During deflation, the lava level at the bottom of Halemaʻumaʻu falls. With less lava, the mountain itself shrinks. The slope of its slopes changes, becoming less steep. Lava refuses to flow from cracks on the surface of the mountain.

Deflation events go on for days at a time. Inflation follows, with the lava level rising and the mountain swelling and lava bubbling to the surface in red-hot springs of liquid rock.

While we wait for inflation to replace deflation, we search for other ways to get close to live lava, to breathe the fumes and feel the heat. But the part of the crater rim that looks down into Halemaʻumaʻu is off-limits. Nongeologists need not apply. The viewing platform has been closed. All we can do is marvel at the caldera and the plume of steam rising up from Halemaʻumaʻu. We consider coming back at night and slipping under the barriers to trespass, but we decide against it, mostly because we are more interested in seeing live lava close up, in touching it, in feeling its heat. At Halemaʻumaʻu all we would see is glowing steam. But we are hindered, too, by self-interest, by risk aversion, by the absence of any reliable information about the danger posed by the steam other than the knowledge that gaseous sulfur oxides mix with water to become sulfuric acid, and a vague feeling that breathing hot sulfuric acid may not be the ideal cure for what remains of our fevers.

✳

We walk down into an adjacent crater, the Kilauea Iki crater, literally the "small spewing" crater. But Kilauea Iki spewed more than a small amount of lava in 1959, converting the crater floor from a lush rain forest into a lava lake four hundred feet deep. It was not the quiet flow typical of shield volcanoes. At times the eruption sent ash nineteen hundred feet into the air, spewing and spitting. At times it spewed for days on end, at other times for just a few seconds. Then it stopped, leaving behind the lava lake with floating islands of cooled lava crust.

A trail takes us across blocks of lava around the edge of the crater, a bathtub ring of basalt rubble left stuck to the crater walls when the lake cooled and collapsed. In the bottom of the crater, we pass steaming cracks. They are not sulfurous. The steam here is from groundwater, rain draining into the ground and encountering still-hot rock, rock that, though not molten since sometime before 1995, remains capable of boiling water.

I put my hand deep into a crack, and a sudden burp gives me a near scalding. I decide that I will no longer reach into the darkness of steaming cracks.

I wander off the trail to find holes drilled into the hardened lake. They were drilled in 1960, 1961, 1962, 1967, 1975, 1976, 1979, 1981, and 1988. In September 1960, the rock six feet below the surface was molten and nearly two thousand degrees. By 1962, drillers had to reach forty-two feet to find liquid rock. By 1981, they had to reach 190 feet. By the 1990s, after the drilling program ended, any remaining molten rock was trapped in tiny, scattered pockets. But the hardened rock, underground, remained incandescent, glowing red.

Now it would be difficult to find glowing rock. Glowing rock may no longer be present. But the mathematics of heat transfer and cooling suggest that earthen fevers of several hundred degrees remain alive and well.

Ferns have taken hold along the edges of steam vents. ʻŌhiʻa lehua shrubs grow here and there. There are ʻōhelo ʻai shrubs, and kūpaoa, and pilo, and kopa. But we walk on rock. Plants have colonized the crater floor in the same way that they colonize sidewalks. That is, they have colonized cracks. And they began colonizing cracks by 1962, when the crater floor was clouded in acidic steam. First came the ferns, but the shrubs followed within two years, working their way inward from the edges of the crater.

We walk to the vent itself, the opening from which lava, ash, gas, and heat once spewed. Now it lies cool and disappointing, not even a considerable hole in the ground, only a depression full of rock. Twain, had he been alive to see this, would not have been impressed. I sit on the edge of the depression and eat a candy bar. Small raindrops fall. A rainbow appears, arcing across the crater floor, its ends touching the crater's walls.

In the rain, I think of Twain and Tyndall. In 1863, just three years before Twain sailed for Hawaii, John Tyndall wrote about the temperature of the earth, explaining why the planet was as warm as it was. "Aqueous vapour is a blanket more necessary to the vegetable life of England than clothing is to man," Tyndall wrote. "Remove for a single summer-night the aqueous vapour from the air which overspreads this country, and you would assuredly destroy every plant capable of being destroyed by a freezing temperature. The warmth of our fields and gardens would pour itself unrequited into space, and the sun would rise upon an island held fast in the iron grip of frost."

The realities of atmospheric physics do not change the reality of cold water on still feverish skin. We walk out of the crater, under the arc of the rainbow. Immediately above the bathtub ring left by the now dried and shrunken lava lake, we walk beneath a thick canopy of trees. A forest just like this once grew in what is now the basin of Kilauea Iki. That forest was buried under four hundred feet of liquid rock.

Farther away, well out of the crater, we come to a place where Kilauea Iki killed more trees, but here it did the job with ash rather than lava. In places, the ash was forty feet deep. Only the tops of dead trees stood above the ash. Leafless, 'ōhi'a lehua trees were bleached by the sun. I touch one and feel a smooth surface, something like driftwood. They appear dead. But in places there is new growth, fresh branches growing from bleached wood. I look closely. The new growth, I know, is not from seeds that have fallen on these bleached trees. It is from the trees themselves. Ten years after being buried under hot ash, the tops of certain 'ōhi'a lehua trees sprouted new growth, thumbing their noses at those who thought they were dead, thumbing their noses at the goddess Pele.

✳

Kilauea inflates. From space, the lava looks promising. The warm spots seen by yesterday's satellites have grown into hot spots. "But they're a long way up," my geologist tells me. "Four and half miles as the crow flies."

We are not crows. We have to walk up Kilauea, four and a half miles as the crow flies, or six miles as the writer walks, up a trailless slope of recently hardened lava. We work for our scenery.

These slopes were not always trailless. They once sported not only trails but roads. And houses stood along those roads, looking down on the Pacific. But in 1990 lava came down Royal Avenue and Princess Avenue, flowing toward the sea. Traveling at a pace slower than a slowly walking man, the lava did not stop for stop signs. It overwhelmed speed limit signs. It flooded houses. It entombed an entire subdivision. That subdivision, still present beneath hardened black rock, was called Royal Gardens.

We park at the end of the access road to Royal Gardens, at a point where lava flowed across the road, drying and igniting wet brush along the way, burning the grass along the road's shoulder and finally burning the asphalt itself. We strap on packs full of water. I also carry my thermometer and a pan of Jiffy Pop popcorn.

We walk around mounds of boot-eating 'a'a lava. We tread on the smooth, hardened pillows and ropes and intestines of pahoehoe lava. In places, the way forward looks like a hardened whitewater river, frozen and blackened. We walk up and down across big standing waves and step on smaller ripples and cross main currents and back eddies. This rock hardened years ago, yet steam still rises from cracks.

We walk over Royal Gardens. It is beneath our feet, under six separate flows of lava, one on top of the other and all on top of the subdivision, twenty feet below us, or twenty-five feet. Under the lava is a treasure trove of fire artifacts, of metal toys and silverware and nails and iron plumbing, with the plastic and wood and rubber and glass burned or melted away.

Ahead of us we hope to see thick steam, but instead we see only clear skies and shadeless black rock under a glaring sun. The uphill climb is hot work. I lag behind, letting the geologist and my companion move forward without me. I drink water, fighting off what is left of my fever.

I search, with little hope of success, for the chirpless cricket, *Caconemobius fori,* an insect known only from the unvegetated lava flows of Hawaii. The chirpless cricket is the fresh basalt equivalent of the fire beetle. It was for some time assumed that animals followed plants in the colonization of new lava, but in fact the cricket comes first, moving onto the lava to feed on whatever the wind provides, hiding in crevices during the day and foraging at night, chirpless throughout to avoid attracting predators. In 1969, 153 silent crickets were captured during six days of trapping, all of them well away from the edge of the flow, well away from any sign of vegetation, on terrain similar to this. But I have neither traps nor time. I push ahead.

It is two quarts of water and another hour to the crest of the hill, to where we expect to see live lava. On this part of the mountain there is no steam. Nor is there smoke. But in places heat shimmer cloaks the horizon, turning sharp outlines to vague, wavering shapes. It is the same heat shimmer that appears above hot ground to create mirage lakes in the desert near Las Vegas. Hot ground heats the air. The hot air expands, becoming less dense, leaving more space between molecules. It rises, forming chaotic eddies. Light travels fastest where the air is hottest, and each time light changes speed, it bends. It creates a shimmer.

Hot air over hot rocks would show up on satellite images. Some of it could be heated by the sun, but geological heat is at play here. We walk on rock that has recently been molten, some of it no more than a few weeks old. We walk on a solid shell above incandescent rock. Heat finds its way up my pants legs to the bare skin of my calves and thighs.

Now the air smells. It smells of sulfur but also of something organic, of drying vegetation or burning wet leaves.

I pull my thermometer from my pack. Under my feet, the ground offers 130 degrees. This is not Death Valley. The sun, acting alone in this part of Hawaii, would not warm the ground to 130 degrees.

Things are looking up.

We make our way toward an island of trees covering a third of an acre. They are optimistic trees, growing on ground that was liquid only decades earlier and surrounded by ground that was liquid within the past few months. I dub the tree island

Fort Apache. Heat shimmer shrouds the left and right of Fort Apache. Steam rises from the right edge of Fort Apache.

A Fort Apache shrub bursts into flame.

This is it: live lava. We can see it now, close to the burning shrub. Little tongues of liquid earth make their way along the edge of Fort Apache. Through binoculars, I watch glassy gray ground suddenly crack, and the cracks glow red, and lava emerges, flowing with a consistency something like that of melted chocolate. The lava advances an inch every one or two seconds.

We move in. From ten feet away, I point my thermometer at flowing lava and read a temperature of 1,560 degrees. In fact, it is probably hotter. My thermometer does not provide accurate readings at temperatures this high. From here, I feel as if I am standing next to an open furnace. And, in fact, that is exactly what I am doing.

We move closer.

Inches behind the molten red tongue of fluid, the surface fades from glowing red to the color of brushed mercury, a dull metallic shine. It pops, sending tiny shards of rock upward as the surface cools and contracts. The shards cut fine arcs two feet into the air. I wrap a shirt around my face and don leather gloves. I step closer. To stumble forward would be to stumble into liquid rock. With a geologist's hammer in hand— a little pick the size of a carpenter's hammer but with a metal handle—I stretch my arm toward the source of heat. I reach into the lava.

Moving the hammer through lava is something like moving a spoon through cold maple syrup, or like moving a shovel through newly mixed thick concrete, but at the same time like neither one. It is perhaps not like anything other than moving a hammer through liquid rock.

The heat discourages lingering. My face—where my eyes are

exposed, but also under the shirt that I am using as a shroud—feels unbearably hot, close to blistering. My hand—the hand holding the hammer—is burning through the leather glove. I am not sure if it is the heat on my face or on my hand that forces my retreat. Either one would be enough. I step back and cast off my glove, throwing the hot leather to the ground.

"It is possible to dip the hand for a short time into melted lead," Harry Houdini wrote, "or even into melted copper, the moisture of the skin supplying a vapor which prevents direct contact with the molten metal." I have no interest in proving him right and even less interest in proving him wrong. I retrieve my glove and move back to the lava flow, dipping the hammer in again, scooping out a gob of new rock, then letting it cool enough to take the shape of the tip of the hammer, a lava souvenir.

On the side of the mountain, as the glowing red tongue of lava extends outward, the hardening rock behind it swells and contracts, as if alive. The breathing contortions occur in the freshly cooled rock with the color of brushed mercury, but also in the rock farther back, rock that has faded to a duller shine, and even farther back, in the black rock behind it that has been cooling for several minutes and now looks hard and brittle but is not.

In places, the rock swells to the point of cracking, and the cracks glow.

We move away, working our way around Fort Apache, looking for more flows, for more of what the geologist calls "breakouts." He moves straight across ground recently solidified, ground that had been liquid within the past hour. In places, the ground flexes beneath his feet. I take a longer route on cooler ground.

"You just have to keep moving," he calls, "to keep your boots from melting."

I follow him. The tip of my walking stick, made of soft rubber, bursts into flame. I snuff it out against hot young rocks. I keep moving. My boots do not melt.

We find a breakout with multiple tongues of fresh lava. One tongue is moving quicker than the others, forming a red stream, its outer edges hardening to form levees. The levees contain the tongue, preventing it from spreading sideways and sending it forward. The tongue flows like a stream over older lava, ground already hardened and cooled. Scattered ferns growing in the old lava are overwhelmed one by one. They steam for a moment and then flame, surprisingly large and long-burning flares at the lava front, releasing a smell of baked fern. The lava finds an 'ōhi'a lehua shrub, surrounding its trunk, riding up the wood to form a tube several inches tall. Its leaves steam and then flame, the bottom leaves first. Ultimately all the leaves ignite. The trunk and its branches steam but persist for several minutes. The lava around the trunk turns black, forming a mold before the wood itself finally burns.

We stand between two tongues of lava. A breeze blows from behind us, toward the lava, keeping its heat at bay, but when the breeze dies for a moment, the heat confronts us suddenly, reminding us of where we are, reminding us that we are standing next to, and on top of, a furnace.

With the burned tip of my walking stick, I poke at young black lava. It looks like rock, but the surface bends inward like soft plastic, flexible but impossible to penetrate.

I take my pan of Jiffy Pop from my pack and put it on top of freshly hardened lava. The heat prevents me from staying close to the pan, from keeping the pan moving. Within seconds the butter inside sizzles and the kernels pop, but without motion, the kernels on the bottom burn. The smell of burned popcorn replaces the smell of baked ferns.

Late in the afternoon, we head downward. Crossing newly hardened lava, we keep moving, and soon we are on older lava, on ground that hardened a year ago and two years ago and ten years ago. After a few minutes, we share the last of the water, looking back toward the flows. From here the live flows are invisible, hidden behind the dark line that marks the crest of the slope. We move on, downward, following the path of the lava, walking on young earth.

My fever is gone, no longer noticeable, my febrile search for live lava behind me.

Chapter 7

BOOM

Had the government prevailed in the early 1960s, temperatures just below the tundra surface of northwest Alaska's Cape Thompson would have jumped from somewhere around freezing to something like twenty million degrees. Atoms of heavy hydrogen, a key ingredient in thermonuclear weapons, would have collided and fused. In fusing, a portion of their mass would have been converted to energy. For a moment, a dot of ground in the Alaskan hinterlands would have been hotter than the surface of the sun. The double flash of light characteristic of hydrogen bombs would have been hidden beneath the tundra, but its heat would have abruptly vaporized soil and frozen groundwater. The blast would have sent seventy billion pounds of frozen dirt upward and outward, along with assorted animals and plants. Lemmings and ground squirrels would have flown through the air, and perhaps a surprised grizzly bear and foxes and maybe a few musk oxen, all sent skyward along with tundra grasses and flowers and the tiny shrubs that grow on windswept, frozen ground. It would have made the Sedan shot—the shot that created the Sedan crater in Nevada—look like a firecracker.

Seawater rushing into the blast hole would have boiled instantly. A massive cloud of steam would have risen above the tundra.

If the government were to be trusted, the Inupiat village of Point Hope, thirty-five miles from the blast, would have been safe. In the local language, the village is called Tikigaq, a word that describes the shape of the peninsula on which the village sits, the shape of an index finger curved outward into the Chukchi Sea. Point Hope is known to be among the oldest of the permanently settled Eskimo communities in Alaska. For more than a thousand years, the Inupiat remained here, staying warm enough to survive by burning seal oil and eating whale blubber. Before the Inupiat, another group of people, the little-known and long-gone Ipiutak, inhabited the same patch of land.

Had the government prevailed in the early sixties, a roughly rectangular and somewhat radioactive harbor a half mile wide and three-quarters of a mile long would have been connected to the sea by a channel deep enough for ships. Both harbor and channel would have appeared in a matter of seconds as fission and fusion converted the ground beneath to vapor. Now, had the government prevailed, that harbor would almost certainly be used to stage vessels for oil exploration in the Chukchi Sea.

What the government called Project Chariot, produced and directed by the Atomic Energy Commission, was the brainchild of a postwar program called Plowshare, which itself grew from Eisenhower's Atoms for Peace speech to the United Nations in 1953. The "greatest of destructive forces," Eisenhower told the delegates, "can be developed into a great boon for the benefit of all mankind." Twenty thousand copies of the speech were printed in ten languages. "It is not enough to take this weapon out of the hands of the soldiers," he said. "It must be put into the hands of those who will know how to strip its military casing and adapt it to the arts of peace."

My companion and I book flights and pack our bags.

✳

The road to Point Hope is not a road at all. Point Hope, like many of Alaska's villages, lies well off the road system. It is on the coast, between the Novarupta volcano on the Katmai Peninsula to the south and Barrow to the north. It can be reached by barge or ship or, more easily, by small aircraft from Nome or Kotzebue. We go by small aircraft, a twin engine. We are accompanied by a pilot, a Bible-reading construction worker, two backpacks, a shotgun, twelve cardboard boxes of groceries stacked against a bulkhead, three empty seats in the passenger compartment, and one empty copilot's seat. We fly north along the coast, passing over the village of Kivalina, the thin beaches that separate the village from ocean waves rapidly disappearing, being eaten by a warming and rising sea, its people wrangling with the government for years now, pushing plans to move the entire community to higher ground.

Over the pilot's shoulder, I read gauges that give altitude and fuel levels and temperatures of 200 degrees for both engines, 200 degrees from tiny explosions against pistons and from the friction of moving parts.

It is September and cold here above the Arctic Circle. The cold seeps into the plane, but my feet sweat in heavy socks and boots. To the left a whitecapped Chukchi Sea stretches out toward Russia. To the right, north of Kivalina, a gravel beach marks the shoreline. Behind the beach, lagoons stretch along the coast. Behind the lagoons, rolling hills of tundra reach into the distance. We pass over a long, empty road that leads from a dock to the interior, to the Red Dog Mine, invisible from here but famous as a source of lead and zinc and controversy.

The pilot points out Cape Thompson, the site of the abandoned Project Chariot, the place where Edward Teller would have touched off four hydrogen bombs to make his harbor, but also to test his bombs, to see what two megatons of explosive power could do, to see what happens when you pop off buried devices holding 160 times the explosive power of the bomb dropped on Hiroshima. Teller, genius that he was, knew that there were no funds to do anything but set off the blast and study the amount of dirt that it moved, the radiation it released, and the water-filled crater it left behind. There was no money for wharfs or roads or warehouses. Wharfs and roads and warehouses were not Teller's business.

Years earlier, Teller had worked at Los Alamos on the Manhattan Project under Robert Oppenheimer, "Oppie," as he was known. Teller, it is said, dreamed up the hydrogen bomb while working for Oppie. The hydrogen bomb was dubbed "the Super" to distinguish it from the fission bombs that would soon be dropped on Japan. The Super became Teller's personal obsession.

Teller is seldom described as a patient man. "What Edward can't carry in his head and solve in his head," a colleague once said, "he doesn't want to bother with." He could not carry all the details of a hydrogen bomb in his head. He had help from other geniuses, from machinists, from technicians, from politicians, from taxpayers' dollars. But despite the help, Teller became known as the father of the hydrogen bomb, as if he had sketched the plans at his kitchen table and assembled the Super on a workbench in his garage.

Teller, looking for a place to see what his Super could do, found Cape Thompson on a map, an empty stretch of Arctic coast. Its proximity to Russia—two hundred miles across the

Chukchi Sea—may have been an added attractant. But he was stopped by a handful of academics and the Point Hope Inupiat, hunters who spoke a guttural language and ate whale and seal and caribou and who at that time lived in sod houses. And so the valley next to Cape Thompson lies more or less intact, we hear, but for eroded airstrips and abandoned equipment. Ogotoruk Creek, shallow and clear, sparkling as it falls over stones, winds through tundra as it finds its way down the valley. Next to its mouth, the steep bluffs of Cape Thompson rise vertically from the sea, crowded with puffins that are visible even from our altitude.

When I tell the pilot that we intend to walk from Point Hope to the Chariot site, he looks at me sideways. "You're going to walk?" he says.

"Yes," I tell him. "It is thirty-five miles each way, and we have five days."

❊

The blasting cap is to the stick of dynamite what the atom bomb is to the hydrogen bomb. The blasting cap has little power compared to dynamite, but it creates the temperatures and pressures needed for detonation. The atom bomb has little power compared to the hydrogen bomb, but it creates the temperatures and pressures needed to force hydrogen molecules together, to ignite the fusion reaction that makes the Super super.

Another difference between the atom bomb and the hydrogen bomb: atom bombs have been dropped on cities.

Gokoku Shrine, ground zero at Hiroshima, was next to an army facility. The shrine had been built as a memorial to victims of the Boshin War, a civil war fought in part over the opening of Japan to foreigners in the 1860s, in the time of the Pennsylvania oil boom and Mark Twain and John Tyndall.

After the bomb fell, gravestones one thousand feet from the

Gokoku Shrine were fused with mica. Mica melts at 1,600 degrees. Roof tiles a third of a mile from Gokoku Shrine melted. Clay roof tiles of the kind used on homes in Hiroshima melt at about 2,400 degrees. Telephone poles two miles from the shrine were charred. Telephone poles made from Japanese cedar ignite at about 500 degrees.

The temperature at the shrine itself reached 11,000 degrees, four times hotter than the inside of a blast furnace. Anyone close to the shrine was immediately incinerated. Survivors described birds igniting in midflight, sounding like the men who fight the hottest of forest fires.

The flames near the center of the city sucked air from outside, creating raging winds. The fire and smoke created their own weather. A thick cloud formed above the city, a miasma. From the miasma, rain fell in heavy drops.

Nine out of ten people within a half mile of the shrine died within minutes. Farther out, survivors wandered, injured and in shock, as the firestorm engulfed their city. Men and women staggered with burned hands stretched in front of them, hands that they had held out to protect themselves from the heat, hands now in so much pain that they were held aloft, as if in prayer. Soldiers, staring toward the source of heat, apparently in search of American bombers, were blinded, their eyes burned away. Others, immobilized by their burns, lay on riverbanks and drowned when the tide came in.

This was a time before the possibility of atomic bombs was widely known. The survivors on the ground speculated wildly about the air strike. Only three planes had been seen. The Americans could have dropped *Molotoffano hanakago*—Molotov flower baskets, a kind of cluster bomb. They could have sprayed gasoline. They could have sprinkled magnesium powder onto the city, and the powder, hitting power lines, could have ignited.

On August 7, 1945, the day after the bombing, from a Japanese radio broadcast: "Hiroshima suffered considerable damage as the result of an attack by a few B-29s. It is believed that a new type of bomb was used. The details are being investigated."

One hundred forty thousand people died. I feel sick when I read the history of Hiroshima. I feel sick to write about it.

✳

In the late 1950s, environmentalists did not call themselves environmentalists. There was little that would be recognized as an environmental movement by today's standards. When the government hired the University of Alaska in Fairbanks to undertake environmental studies at Cape Thompson, Teller may have assumed that he was buying friends. Instead he found biologists like Bill Pruitt and Leslie Viereck who openly criticized the government for botching facts and downplaying biological risks. He found a young fisheries ecologist named Tom English who called the Chariot scientists "the firecracker boys" and talked of the Atomic Energy Commission's "mendacity." Teller sewed up the support of pro-development political leaders in Alaska but failed to consider the irritating academics, the Inupiat, and the annoying reality that what he proposed to do was horribly unreasonable.

While the scientists worked at the Project Chariot site, Teller moved forward with other experiments to better understand the usefulness of nuclear weapons as landscaping tools. In Nevada, at ten o'clock in the morning on July 6, 1962, Teller and his boys set off the one-hundred-kiloton Sedan shot and created the rock art crater that my companion and I visited in the Nevada desert. The blast sent earth boiling upward, creating an expanding bulge on the desert floor that grew to eight hundred feet across and three hundred feet tall. Glowing-hot gases burst forth. Bits of desert shot into the sky and outward two thousand feet. A pressure wave rolled along the desert surface to a distance of a half mile. Radioactive dust rose higher and moved faster than expected. By the time the Atomic Energy Commission reassured the public about the blast, a radioactive cloud had spread well beyond the bomb site. Within hours it had crossed two highways, dumping fallout along the way. Road workers were evacuated. Highway 25 was closed. By midafternoon the cloud traveled two hundred miles and blocked out the sun in Ely, Nevada. Within five days the radioactive cloud passed over parts of Utah, Colorado, Wyoming, Nebraska, and the Dakotas. Five days later, tanker trucks and fire engines washed radioactive dust off a seven-mile stretch of Highway 25. Two years later, the Atomic Energy Commission admitted that fallout had probably reached Canada.

Our plane lands on a gravel strip near the old village of Point Hope. A cold, hard wind rips in off the Chukchi Sea, the same sea that in the early 1970s ate this part of the village, flooding homes and swallowing archaeological treasures, forcing people to move several miles east along the coast. People say now that the flooding and erosion were an early expression of climate change, of a warming sea, but in the seventies it was just bad luck. Owners of wooden houses used runners to slide their homes to a new location. Owners of sod houses abandoned them in place and moved into wooden homes on the new town site.

All along the Arctic coast, erosion makes a meal of low spits of sand and feasts on seaside bluffs, taking bites of the coast here and there, slowly changing the map. Warmer temperatures mean less summertime sea ice, and with less sea ice, the water has more time each year to gnaw at the shoreline. Now the late summer winds blow across hundreds of miles of open water, building waves capable of ripping into the shore like a polar bear into the carcass of a whale.

This far north, the soil is often mostly ground ice, water frozen within the dirt, archetypal tundra. Wide gravel and sand beaches line part of the coast, along with scattered rock bluffs, but there are also long stretches where tundra and sea stand in close contact, with the tundra dropping away in bluffs that stand five or ten feet tall. During summer waves lap at the bottom of the bluffs, melting the ice in the soil to dig out sea caves, and the bluffs collapse under their own weight. Tundra bluffs have been collapsing since the end of the last ice age, but in recent years, with the sea ice so obviously disappearing, the soil collapses with startling swiftness. The coastal landscape changes as quickly as it can be mapped.

※

The Scotsman Joseph Black discovered something about ice. Although Black never visited the Arctic, to understand his discovery is to understand the importance of the loss of sea ice.

Black worked during the time of Antoine Lavoisier, and he knew and worked with James Watt. He was trained as a physician and as such knew of fevers, but his passion was chemistry. His scales—very accurate balances for laboratory use—were used by Faraday and Tyndall.

Like these men, Black was known for charismatic lectures. He could be seen extinguishing the flame of a candle with what he called "fixed air." Another name for fixed air: carbon dioxide.

In a spare moment, in a break between building scales and giving lectures and extinguishing candle flames, Black discovered what would become known as latent heat. He discovered that a certain amount of heat—the latent heat of freezing—is needed to break down the organized structure of ice. A piece of very cold ice—a piece of ice at, say, five degrees—warms evenly only until it reaches thirty-two degrees, and then the warming stops until the ice melts. Apply a candle's warmth to a block of ice, and the block of ice steadily warms, but as the ice turns to water, the warming stops. When the ice is gone and only water remains, the water steadily warms.

When the sea ice melts early in the year, when the sea ice disappears, it is expressing more than a steady, gradual warming. It has absorbed enough heat to break down the organized structure of water molecules standing in formation, the hydrogen of one water molecule locked to the oxygen of another.

And when the sea ice disappears, when the sea ice becomes liquid water, something else happens. It loses its reflective surface. It suddenly goes from a hard surface that reflects light and heat to a fluid that absorbs light and heat. It reaches a tipping point, one that is less than welcome to the animals and people who live along the coast and on the sea ice itself.

My companion and I wander around what is left of the old village. Sod houses have been reduced to mounded earth surrounding shallow depressions, rectangular hollows with sparse, windswept vegetation. Even more sod houses are now invisible, swallowed by the warming sea.

Many of the collapsed sod houses were occupied as recently as the 1960s. A few were still occupied in the 1970s. The residents were Inupiat Eskimos, the same people who live here now. But there are older remains too. There are signs of the Ipiutak, the people who predated the Inupiat. What had once been Ipiutak houses can now be described as a patchwork regularity in the tundra surface, an unnatural geometry that speaks of human activity. In the past, these Ipiutak houses might have been described as dugouts covered by sod with a tunnel entrance. There also may have been less permanent structures above ground, shacks made from driftwood or tents made from hides. The village was more than just a few families scraping out an existence. There may have been as many as eight thousand Ipiutak living here at one time. It was the largest settlement known to have existed in Alaska before the arrival of Europeans.

The Ipiutak disappeared around the time that the Inupiat arrived.

I talk to an Inupiat hunter from Point Hope. He tells me that the Ipiutak did not hunt whales. His people did and do. He takes me to the collapsed remains of the sod house where he was born. He lived here until the 1970s, when he moved to a wooden house in the new village, farther from the edge of the hungry sea. An electrical box stands outside the remains of the collapsed walls of the house of his youth. Wires run from the box into what is now a mound of sod.

A single sod house remains standing, nearly intact. This is not the sod house of settlers in the American prairies. It looks more like a hillock than a house. The doorway, framed in the bones of a whale, leans to the left. More bones shore up the walls. Earth carpets the floor. In 1970 a single bare lightbulb hung from wires buried in the sod ceiling, unpretentious but welcome in the Arctic winter. When he was a boy, the hunter tells me, the light of the bare bulb would have been supplemented by the light of a seal oil lamp, a shallow bowl or tray with a chunk of seal blubber and a wick, burning, sending out flickering light and heat against pale jawbones and dark earthen walls. A family of six may have lived here within this mound of sod, cherishing the heat of the seal oil lamp, resting through the winter while outside temperatures hovered near forty below and winds howled in off the sea ice. I cannot imagine myself living in this house.

In Point Hope, heat is life in winter. There is the outer heat of a seal oil lamp and the inner heat from digesting a meal of walrus. There would be the heat of whale oil, too, the same oil that brought the commercial whalers here in the middle of the nineteenth century. Their ships sailed from places like Nantucket and New Bedford and Long Island.

In 1848, Captain Thomas Roys pushed a thousand miles north of the rest of the New England whaling fleet and discovered the whaling grounds north of the Bering Strait. In 1849, ten years before the Drake well, fifty whale ships followed Roy's lead. In 1852 more than two hundred whale ships headed into the northern whaling grounds. They would leave New England in autumn, round Cape Horn, overwinter in Hawaii or San Francisco, and follow the ice north in spring. Over sixty years, something like eighteen thousand whales were killed.

At this time, it was the whale oil that paid the bills. Before the bowhead population collapsed, in a summer season a single whale ship might expect to take ten whales offshore from Point Hope and Point Lay and Barrow and Wainwright and the other coastal villages of the Alaskan Arctic. A bowhead whale could yield more than a hundred barrels of oil. The record, from a whale taken in 1850, was 280 barrels of oil. At the dock in Nantucket, the oil from a single whale might sell for more than five thousand dollars, the equivalent, after a century of inflation, of close to a hundred thousand dollars.

While oil from bowhead whales did not burn as cleanly as oil from sperm whales, it burned well enough. It could also be used as a lubricant for wagon wheels and the inventory of machinery that grew with the Industrial Revolution, a lubricant to control the heat of friction.

The blubber—thick layers of fat held stiff with strings of collagen—was carved into blocks or strips and rendered into oil in onboard refineries, heat separating the good stuff from the bad, the gasoline from the tar, the oil from the blubber.

"Besides her hoisted boats," wrote Herman Melville, "an American whaler is outwardly distinguished by her tryworks."

Melville compared the tryworks to a kiln amidships, supported by thickened beams and held to the deck by steel pins. "The intense heat of the fire," Melville wrote, "is prevented from communicating itself to the deck, by means of a shallow reservoir extending under the entire inclosed surface of the works." Sailors replenished the reservoir's cooling water as it steamed away.

Huge pots sat on top of the tryworks. When not in use, the pots were big enough to hold a napping sailor. In use, fire from the tryworks heated blubber in the trypots. The blubber, tried out, would be scooped from the trypots, leaving only the oil behind. Fresh blubber would be added.

The tried-out blubber, shriveled and hardened, called "fritters" by the whalers, still held enough oil to feed the fire under the trypots. "Like a plethoric burning martyr," Melville wrote, "or a self-consuming misanthrope, once ignited, the whale supplies his own fuel and burns by his own body."

The whales were not stupid. Three years after Captain Roys opened the Bering Strait whaling grounds, one captain wrote that the bowhead was "no longer the slow and sluggish beast we first found him." The whales recognized the sounds of whaleboats. "They don't like the cold iron," wrote a whaler in 1850. A block of ice dropping from the ice pack would slap the water to no effect, but the whales would vanish at the sound of a whaleboat gently bumping the same block of ice.

Within a few years, the commercial whalers wiped out a quarter of Alaska's bowheads. As early as 1851, just three years after Captain Roys had opened the Bering Strait whaling grounds, one captain wrote, "Where I whaled last voyage, now looks like a deserted village." A year later, the first mate of the whale ship *Montreal* wrote, "No whales, the ground appears as barren as the deserts of Arabia, altho we are on the very spot where last May we saw whales in abundance." Within twenty years, half the bowheads were gone, rendered into light and lubricant and heat for the cities and towns and farms of nineteenth-century America.

As the number of bowheads diminished and whaling became more challenging, the whalers realized that they could overwinter in the Arctic. They could hunt in early spring when the whales migrated through open leads in the melting ice. By the 1880s, the New England whalers built their own sod houses just east of Point Hope in a place known to the Inupiaq as Quzmiarzuq and to the whalers as Jabbertown, after the many languages spoken there by men who had signed on from scattered ports.

By now the people of Point Hope were dying as quickly as their whales, killed by syphilis, smallpox, influenza, measles, diphtheria, and tuberculosis, all compounded by alcohol. Families lay in their sod houses, mother and father and son and daughter side by side on hides of polar bear and caribou, coughing, fading into the delirium of deadly fevers, sweating while a winter wind blew above.

Like their whales, half the Inupiat population of Point Hope died in a single generation.

Atom bombs, for all their suddenness, were not unique in their ability to incinerate human beings en masse. Firebombing was widely used in Germany during World War II. The idea was simple: use aerial bombing to ignite a firestorm, to convert a city into a furnace, to cripple the war machine that supported the troops, to destroy not only factories but the houses of workers. For a time, firebombing missions were called dehousing missions. Hamburg was among the first cities to be dehoused. For some time afterward, to dehouse a city was also to Hamburgize the city. Essen, Duisburg, Dusseldorf, and Hanover were Hamburgized. In Japan, Osaka and Kobe and Tokyo and Okayama and sixty-three other cities were Hamburgized.

Napalm—jellied gasoline, a thickened version of gasoline that stuck to whatever it landed on and burned ferociously—was dropped from the air. A single bomb could envelop an area 270 feet long and 80 feet wide in flames burning at 1,500 degrees.

From Robert Haney, an American prisoner of war held in Osaka: "On the night of March 13, 1945, Osaka was bombed. Our camp was barely one city block inland. The first firebombs hit about two blocks inland and continued away from us four or five miles. The raid lasted much of the night. In the morning, a vast area—later determined to be 25 square miles—was a smoldering desert."

Obata Masatake, a survivor of the Tokyo firebombing, later talked to a radio interviewer. "There were some old women wearing those thick quilted coats with padded hoods," he said. "But they were getting so hot they pushed them back without thinking. It was fatal. It was so hot their hair just burst into flames, just like that and there was nothing I could do to help. It was so hot I couldn't breathe."

After a firebombing, charred bodies lay unattended in the streets, blackened corpses, mummified by heat, without hair, partially clothed or with clothes burned entirely away. Smoke and steam rose from the streets around the corpses.

Firebombing raids typically killed tens of thousands of men, women, and children. Many died from the burns themselves. Others suffocated as flames sucked oxygen from the air. Others were trampled during the panic that accompanied fire bombings.

The American aviator Ray "Hap" Halloran was shot down over Tokyo. He watched B-29s flying over the city. He watched the sky turn to flames. "I prayed for myself," he said, "and I also prayed for them, too."

Kurt Vonnegut survived the Dresden firebombing and described it in *Slaughterhouse-Five*. In the book, a bird calls, "Poo-tee-weet?" It is as if the bird is trying to make sense of something that can never make sense. The bird is asking, "Why?"

My companion and I walk with loaded packs along the beach. Jabbertown is behind us, modern-day Point Hope is behind Jabbertown, and the abandoned old village of Point Hope is behind modern-day Point Hope. The Chukchi Sea stretches out to the right, and a long brackish lagoon sits to the left. In front of us, the beach runs to the horizon, and beyond that the bluffs of Cape Thompson separate us from what is left of Project Chariot. The wind from the Chukchi blows a light gale. My boots sink into the gravel beach with every step. My feet rub against the innards of my boots, raising hot spots and blisters of the sort sometimes associated with firewalking. I sweat with the fever of activity.

On my back I carry warm clothes, matches, chemical heat packs, a pump-up cookstove that runs on regular unleaded gasoline, a pint of gasoline in a metal bottle, high-calorie dehydrated food in a bear-proof container, a tent, a sleeping bag, a sleeping mat, extra clothes, and a book called *The Nuclear Family Vacation*. Over my left shoulder I carry my shotgun as protection against polar bears and grizzlies, both common along this coast. If needed, I can pull a trigger, activating a chemical reaction that will convert a solid into a high-temperature gas, and that gas, expanding, will push a slug of lead outward, with luck into the chest of a marauding bear before it successfully marauds.

✳

Teller, the man behind Chariot, was in the nuclear bomb game from the beginning. He was among the physicists who convinced Einstein to warn Roosevelt that atomic bombs were a theoretical possibility. Einstein wrote his warning letter to Roosevelt in 1939. "A single bomb of this type," Einstein wrote, "carried by boat and exploded in a port, might very well destroy the whole port together with some of the surrounding territory."

Teller feared both the Nazis and the Soviets. He was born into a middle-class Jewish family in Budapest and grew up in an atmosphere of Nazi and Communist rhetoric and violence. He left in 1926 to study physics in Germany but fled the Nazis in 1933, eventually landing in the United States. His friends Lev Landau and Laszlo Tisza described to him the Soviets' Great Terror of the 1930s, the killing of dedicated scientists, the arrests and assassinations of anyone potentially threatening to Stalin.

Many years later, Andrei Sakharov, the Soviet physicist and bomb maker who eventually fought weapons proliferation and was awarded a Nobel Peace Prize, would say that Teller's "anti-Soviet paranoia" had some basis in reality. Fear of the Soviet regime, Sakharov would say, was a reasonable fear.

On the Manhattan Project, Teller lived in a world of secrecy and compartmentalization. The address for the laboratory at Los Alamos, New Mexico, was a post office box. Fake names were used. Mail was censored. Telephone calls were monitored and interrupted. Nearby communities speculated on the laboratory's purpose. The laboratory sent people to the La Fonda Bar in Santa Fe to spread misinformation about the manufacture of electric rockets. They were building spaceships, making poison gas, building submarines, sheltering pregnant servicewomen.

For scientists accustomed to a world of open discourse, compartmentalization and severe restrictions on communication did not come easily. It was joked that a scientist supervising two departments needed a special security clearance to talk to himself.

Teller was present at the Trinity test, the first explosion of an atomic bomb, in the desert of New Mexico. "I was looking," Teller later said, "contrary to regulations, straight at the bomb. I put on welding glasses, suntan lotion, and gloves. I looked the beast in the eye, and I was impressed."

While his colleagues were inventing the fission bombs that would be dropped on Hiroshima and Nagasaki, Teller was thinking about the Super, the hydrogen bomb, a far more powerful device. The concept was simple. The atom bomb generates heat by splitting atoms, by making elements with large nuclei into elements with smaller nuclei. The hydrogen bomb generates heat by joining atoms, by forcing the smallest of nuclei together, by slamming hydrogen into hydrogen with such force that it becomes helium. Teller would use the heat and pressure generated by the atomic bomb's fission reaction to ignite a fusion reaction, making hydrogen into helium and releasing vast quantities of energy. Fission was kid stuff, while fusion was the source of heat that drove the sun itself.

From the physicist Herbert York, who would become the first director of the Lawrence Livermore National Laboratory in California: "I have this vivid memory of Teller going to the blackboard and just with a few strokes drawing a cartoon that was: 'This is how you make a hydrogen bomb.' And I remember, either at the time or that evening, getting a little bit of the shivers because I realized: That was it."

The journey from conceptualization to explosion required substantial effort. Many of the physicists involved with the Manhattan Project were horrified by the fruits of their labor. They did not want to see an even more powerful bomb in the hands of politicians. Teller pushed it through.

The first Super, named Mike, was two stories tall and weighed sixty-five tons. It was shipped to the island of Elugelab in the Marshall Islands. Observers were told that they would soon see the world's most powerful explosion.

Mike sent out a fireball three and a half miles wide, left a crater a mile across and almost two hundred feet deep, and turned millions of gallons of seawater into steam. When the steam dissipated, the island of Elugelab was no more.

From the physicist Harold Agnew, who was watching from a ship twenty-five miles away from the blast: "Something I'll never forget was the heat. Not the blast. The heat just kept coming, just kept coming on and on and on. And it was really scary. It's really quite a terrifying experience because the heat doesn't go off. On kiloton shots it's a flash and it's over, but on those big shots it's really terrifying."

Years later Teller met Mikhail Gorbachev, and Gorbachev refused to shake his hand.

Walking along the beach with heavy packs, we come across the bloated carcasses of a dozen walruses. They have been shot and beheaded, their tusks harvested to be converted to native art. The people of Point Hope eat walrus, but when a walrus is shot, it sinks. If a hunter cannot get a harpoon or a hook into the walrus before it sinks beyond reach, the meat is lost. But warm-bloodedness dooms carcasses to rapid decay. Underwater, the heat within the now dead sunken walrus promotes bacterial growth. The bacteria—the chemical reactions of bacteria breaking down walrus flesh—generate more heat.

Rotting carcasses of sunken dead walruses eventually fill with warm gas and float to the surface. The beach east of Point Hope tends to collect flotsam from all parts of the Chukchi Sea. These dead walruses could have been shot hundreds of miles away. By the time they land on the beach, their meat is useless, but their tusks, carved, can be worth thousands of dollars. Sometimes the tusks are elaborately carved but left intact within the jawbone, and the jawbone might remain attached to the skull. Skull and jaw and carved tusks are mounted and sold in art galleries in Anchorage, New York, and Boston.

Walruses, fat in life, are swollen in death. But even headless, beached, and swollen, they remain impressive. I kneel next to one. Gale-force winds are not enough to sweep away the stench. The animal is so bloated that I fear it might explode.

In medieval warfare, catapults of various designs were sometimes used to fling the bloated carcasses of animals and the diseased carcasses of humans over the walls of besieged castles. Catapults also threw incendiary weapons.

There were different kinds of incendiary weapons. Liquid fire—also known as sea fire, Roman fire, Greek fire, and war fire—was invented in the late seventh century, possibly by a man named Kallinikos from the city of Heliopolis, "sun city." Later, the emperor Constantine Porphyrogennetos wrote that the secret of liquid fire was "shown and revealed by an angel to the great and holy first Christian emperor Constantine." That secret was closely guarded, but it may have involved the use of sap from pine trees and sulfur, or quicklime, or saltpeter, or calcium phosphide, which, when mixed with water, releases the highly combustible phosphine gas. Or it may have been nothing more or less than petroleum harvested from seeps, an early form of napalm.

Liquid fire could be sprayed from the bow of a ship. From the *Alexiad*, written around 1150: "On the prow of each ship he had a head fixed of a lion or other land-animal, made in brass or iron with the mouth open and then gilded over, so that their mere aspect was terrifying. And the fire which was to be directed against the enemy through tubes he made to pass through the mouths of the beasts, so that it seemed as if the lions and the other similar monsters were vomiting the fire."

From a thirteenth-century memoir: "My opinion and advice therefore is: that every time they hurl the fire at us, we go down on our elbows and knees, and beseech Our Lord to save us from this danger. This was the fashion of the Greek fire: it came on as broad in front as a vinegar cask, and the tail of fire that trailed behind it was as big as a great spear; and it made such a noise as it came, that it sounded like the thunder of heaven. It looked like a dragon flying through the air. Such a bright light did it cast, that one could see all over the camp as though it were day, by reason of the great mass of fire, and the brilliance of the light that it shed."

The memoir spoke, too, of praying, of weeping: "Oh! fair Lord God, protect my people!"

Almost as soon as nuclear weapons were tested, discussion of test ban treaties began. Teller, fearing a cessation of weapons testing, saw Eisenhower's Atoms for Peace speech and the Atomic Energy Commission's Plowshare program as invitations to continue setting off bombs. Teller and his colleagues talked of "geographical engineering." They talked of blasting a sea-level canal across Panama to end the need for pesky locks. They talked of blasting an alternative route to the Suez Canal. They discussed the use of nuclear explosions for mining and oil exploration, for cutting through pack ice, for propelling rockets. A project called Orion assessed the possibility of propelling a sixteen-story-tall spacecraft with nuclear explosions. A team calculated the number of blasts that would be required to fly Chicago through space.

At a press conference in Juneau, Alaska's capital, Teller unveiled Chariot. "We looked at the whole world," he said, "almost the whole world, and tried to pick a spot where we could most effectively demonstrate the peaceful uses of energy." He compared the risk of fallout from a Chariot blast to the risk of radiation poisoning from his watch. He claimed that he could dig a harbor in the shape of a polar bear, if that is what the people wanted. His bombs could do the work of a thousand men with a hundred bulldozers and excavators and dump trucks working around the clock for a year. His plan was to move seventy million cubic yards of frozen earth, and to do it in the blink of an eye.

From the book of Isaiah: "And they shall beat the swords into plowshares, and their spears into pruning hooks; nation shall not lift up sword against nation, neither shall they learn war any more."

From Edward Teller: "If your mountain is not in the right place, drop us a card."

※

Bear tracks decorate the beach on which we walk. I daydream, envisioning my own tracks, through hot ash, footprints of a firewalker. Late in the afternoon, we set up camp on a point of land that juts back into the brackish lagoon behind the beach, well clear of the bear tracks. The sun rides low out over the Chukchi Sea. The wind blows. I pump our cookstove, pushing and pulling on the small silver handle that slides in and out of a pint-sized tank filled with regular unleaded gasoline.

A ship's surgeon named John Simpson, traveling near here around the time of Captain Roys, when contact between Inupiat and Europeans was becoming commonplace, described the means of making fire used by the Inupiat. "For procuring fire," he wrote, "the flint and steel is used in the North, and kept in a little bag hanging round the neck; and in Kotzebue Sound the pipe bag contains two pieces of dry wood, with a small bow for rotating the one rapidly while firmly pressed against the other until fire is produced. In the absence of these, two lumps of iron pyrites are used to strike fire upon tinder, made by rubbing the down taken from the seeds of plants with charcoal."

Again and again, the Chukchi Sea's wind blows out my matches, so I move closer to the stove. Ten matches later, I strike one close enough to get ignition. The stove flares with a momentary vengeance, a miniature fireball that singes the hairs on my right hand and burns the tips of two fingers. But now the stove roars, its flame blue and orange and yellow. The blue bit burns cooler than the yellow, the dark bit in the middle of the flame is hotter than the blue bit, and the flickering halo that surrounds the whole thing burns hottest of all. The heat moves through the metal parts by conduction, is carried away through the air by convection, and warms everything around it by radiating outward.

I heat water on top of my stove while I read the only book in my pack, Hodge and Weinberger's *A Nuclear Family Vacation*. Nuclear test sites and missile silos and bomb laboratories, the authors contend, can be tourist attractions. The authors visit Frenchman Flat at the Nevada Test Site. They tour a Missile Alert Facility in Wyoming. They take in the Kurchatov Institute in Russia and a uranium conversion facility in Iran. They explore the Lawrence Livermore National Laboratory in California. Some of their destinations are located near the homes of relatives: "a cousin in Los Alamos, birthplace of the atomic bomb, an aunt in Las Vegas, near the site where nuclear weapons were once set off on a regular basis, and a brother in northern California, not far from the lab that physicist Edward Teller built." The authors talk to people who are designing bombs, disassembling bombs, and guarding bombs.

In their introduction, the authors ask a simple question: "How does one plan a nuclear vacation?" The answer: one balances the value of visiting a place with the ease of getting there and the likelihood of gaining access. North Korea is scratched off the list. Because of security restrictions, "India and Pakistan have little to offer the nuclear tourist." Point Hope, Cape Thompson, and Chariot do not even make it into the book's index.

The water boils. I prepare noodles. The sun sets. My companion and I eat and then slither into sleeping bags. Wind whips the fabric of our tent, and my burned fingertips throb. The stove's sudden flame had left hot spots, blisters. "It's a mind-set," the instructor at the Firewalking Institute had told me. "Tell yourself you'll get hurt and you will."

Atom bombs fueled by highly enriched uranium are famously simple. Take two blocks of highly enriched uranium, each about fifty pounds. Get a pipe about ten feet long. Shove the first brick of uranium in one end of the pipe and the second brick of uranium in the other end. Pack in high explosives behind the two bricks. Insert blasting caps into the high explosives. Plug the ends of the pipe. Apply enough current to trigger the blasting caps and, hence, the high explosives. Hot gases released by the explosion shove the two bricks of uranium toward one another. They collide, and in colliding they reach critical mass.

Uranium atoms, by their nature, occasionally fall apart. When they fall apart, they send neutrons flying outward. At critical mass, the loose neutrons fly into other atoms, forcing those atoms to break apart. With neutrons flying everywhere, smashing into atoms to set free more neutrons, and more, and still more, a chain reaction leaves in its wake the pieces of broken uranium atoms.

The broken atoms, in breaking, lose mass. It is as if the two halves of a broken cookie weigh less than the whole cookie. The missing mass has become energy: $E = MC^2$.

In this sort of chain reaction, resulting from slamming together two bricks of uranium, the energy released is sudden and devastating. The best thing that can be said about it is that it blows itself apart very quickly, the force of the explosion sending whole and broken atoms of uranium in all directions, reducing the mass of uranium below criticality, snuffing itself out.

Making an atom bomb is simple in principle, but in reality there are tough challenges, especially if it has never been done before. Key among these is the challenge of obtaining the right fissile materials, materials like highly enriched uranium or plutonium. And although simple in principle, there are details to be worked out. In wartime, progress on the bomb required secret cities. It required Site X in Oak Ridge, Tennessee, to enrich uranium, Site W in Hanford, Washington, to produce plutonium, and Site Y at Los Alamos, New Mexico, to work on the details of bomb design. And Teller was already dreaming of his Super, his hydrogen bomb, a device that would require a yet-to-be-invented atomic bomb to set it off, to trigger the fusion reaction that would create far more energy than the slamming together of bricks of enriched uranium.

From start to finish, from the initiation of the Manhattan Project on December 6, 1941, to the detonation of the world's first hydrogen bomb, took eleven years. And Teller was itching to dig an instant harbor along the coast of Alaska.

On day two we walk another eight miles on the gravel beach, beautiful but eventually monotonous. I talk to Point Hope Inupiat hunters out scouting for caribou. They ride Honda four-wheelers, using the beach like a highway to get to the hills surrounding Cape Thompson. It is early for caribou, they tell me, but the migration will arrive soon.

We come across a thirty-foot bowhead whale, dead and floating in the surf, as bloated as a beached walrus. Without dismounting from his four-wheeler, a hunter tells me that the whale might have been killed by hunters from another village. "Maybe Wales," he says, "struck and lost." Wales is well to the south, on a tip of land that forms the choke point of the Bering Sea. The wounded and eventually dead animal would have had to swim or float two hundred miles before washing up here.

Another Inupiat hunter tells me that the sea ice, in spring, melts from beneath. The currents are changing, he says. The elders have never seen ice like this before, ice that melts early and does not return until late in the year.

Ringed seal mothers carve lairs in the sea ice and the overlying snow. In the lairs, they birth pups, small white butterballs, unable to swim for their first few days of life. When the thaw comes early, the lairs collapse. Gulls attack the pups from above.

By midafternoon we come to bluffs that mark the end of the beach, the northern edge of Cape Thompson. The bluff forces us inland, into the foothills. We walk over dry tundra with ankle-breaking tussocks, tight clumps of sedge that form little mounds. We step over a stream that has cut its way into the frozen ground, forming a miniature valley. The hills in this country—the last of the mighty Brooks Range that crosses northern Alaska—are rounded. Pockets of deep snow sit in the shadows on north-facing slopes. Windswept hillsides, sparsely vegetated, look salty from a distance. Weather has exposed pale mineral soils and rock.

Eager to see Chariot, we walk late into the evening, but the remaining distance is uncertain. My companion, wisely, wants to stop. We camp that night on a bed of moss. I pump the stove, holding the pump knob between still-burned fingers to pressurize the fuel tank. I create a flame of gasoline that burns yellow and blue and makes the stove hiss like a jet engine.

The ingenious little stove was developed in response to the U.S. Army's World War II request for something that could run on gasoline or kerosene, something that could be easily ignited in temperatures from 60 below to 125 above, about the size of a quart bottle, and no heavier than three pounds. The engineers at Coleman pulled something together over two busy months, drawing from earlier designs of larger field stoves, which in turn had evolved from even earlier designs of kerosene lamps of the sort used in the 1860s to burn rock oil. World War II correspondent Ernie Pyle called the jeep and the Coleman GI pocket stove "the two most useful non-combat pieces of equipment to come out of the war."

William Coleman's company started as a provider of kerosene lamps and ended as a manufacturer of stoves, first for farmers, then for the army, and now for backpackers. The reliability of the Coleman backpacker stove is renowned. They are sometimes described as being "almost bomb proof," which seems, this close to Chariot, remarkably appropriate.

Ship-mounted weapons that spit Greek fire, described in the *Alexiad*, were early flame throwers, but the modern flamethrower did not come into regular use until the trench warfare of World War I and, even more commonly, during World War II. The modern flamethrower was invented in Germany around 1901, where it was called the *Flammenwerfer*. The soldier wears a backpack with two tanks. One tank holds pressurized propellant gas, and the other holds a flammable liquid. The gun itself, connected by a hose to the backpack, consists of a valve and a pilot flame or an ignition coil. Modern flamethrowers can throw flames more than 250 feet.

Soldiers operating flamethrowers, if taken prisoner during World War I or World War II or in Korea, were often executed on the spot.

Poo-tee-weet.

✻

Across the water, the Soviets had their own Plowshare program. They called it, in English, "Program Number 7: Nuclear Explosions for the National Economy."

The Soviets built an instant though somewhat radioactive reservoir in Kazakhstan. They tried but eventually abandoned attempts to explosively dig a forty-mile stretch of a canal that would have diverted water from the Pechora River to the Kama River, and from there into the Volga River, and ultimately into the Caspian Sea. They built underground storage caverns for oil and gas.

Have an oil or gas well that fails to produce? No problem. Set off a small nuclear weapon near the bottom of the well to fracture the rock, and the oil or gas flows. This was done nine times in the Soviet Union.

Have a gas well that is burning out of control? Again, no problem. A properly designed bomb placed the right distance from the well snuffs out the flame and seals the holes through which the gas found its way to the surface. This was done four times in the Soviet Union. The first time it was tried, it extinguished a well fire that had been burning for three years, a well fire that had defied all conventional means of control and wasted enough gas to power a city the size of Saint Petersburg for several years.

Need to better understand the country's geology? Between 1971 and 1984, the Soviet firecracker boys used thirty-nine atomic bombs for something called "deep seismic sounding," the exploration of the earth's crust and mantle through reflections of shock waves along thousands of miles of transects.

In Anchorage, Teller once told a friendly audience, "The Soviets are doing it; we are not." Before the Soviet Nuclear Explosions for the National Economy program was over, they had set off well over a hundred bombs.

In 1990 the International Chetek Corporation of Moscow tried to privatize the Soviet experience with bombs for peace. At a scientific meeting in Ottawa, Canada, the company's president told the audience that his organization was ready to demonstrate the use of nuclear explosives for the destruction of toxic waste. The idea was simple: haul your waste to a Siberian test facility, inject it a half mile underground, and incinerate it with the extremely high temperatures of a nuclear explosion. The impression of the meeting's chairman: "Everybody in the room thought they were nuts."

※

Now, halfway through day three and still some distance from Chariot, we walk along a winding stream laced with morning ice, the mud and sand of its banks peppered with tracks from caribou and musk ox and bear. My feet are wet.

The stream is a tributary to Ogotoruk Creek, which, had Teller set off his bombs, would now flow into a radioactive harbor. *Ogotoruk,* in the language of the Inupiat, means a hunting bag, a poke, suggesting the importance of the area for subsistence hunters. Had Teller prevailed, we would be walking toward a point that would momentarily have been among the hottest spots in the solar system.

The authors of *A Nuclear Family Vacation* dodged from here to there, looking at sites loosely associated with potential Armageddon. I am not a lover of warmth, I am not a thermophile, but at this particular moment I want to go somewhere that does not involve cold, wet feet and frozen ground. At this particular moment, standing next to the iced-over Ogotoruk Creek, fire-walking sounds increasingly attractive.

We move through stands of chest-high willow, fine cover for grizzly bears. I slide my gun from my shoulder and carry it in my hands. It is foolish to walk through a willow thicket in bear country, but there is no obvious way around.

We break out of the willow and come to a sharp bluff. Just beyond the bluff, the Chukchi Sea comes into sight, dark blue against a clear sky. Where the land meets the sea, we see for the first time the huts and scattered equipment left behind by Project Chariot.

It is another two miles through marshy tussocks crunchy with ice. As we move closer, the wind coming off the Chukchi Sea increases into a gale, and details of the Chariot camp become more apparent. Two huts remain standing, and next to them stands rusting machinery—a grader, a dozer, and an all-terrain vehicle called a Weasel. The abandoned airstrip, too eroded for use, runs inland from the coast.

I expected the Chariot camp to be vacant, empty but for the two of us. In fact there are three Inupiat men working at the camp. "You're lost," one jokes, pointing northwest. "Point Hope is that way." They came in on four-wheelers and have been here for the past week, sent by their village to clean out one of the Chariot shacks, to convert it to a search-and-rescue camp, stocked with food and propane and a radio. They are living on cots in the hut.

The three men have trained the resident ground squirrels to jump up and snatch cookies from their hands. One man lets a squirrel take a cookie held loosely in his mouth. The squirrels are obese, rounded. They look like jumping balls of fur. The men named one of the squirrels Cookie and another Scarface. Cookie and Scarface squabble with loud squeaks and surprisingly vicious mouth-to-mouth combat before running to their respective home turfs.

One of the men asks if I work for the government. Another says that his grandfather's sod house was on this beach, and the family had tried, unsuccessfully, to claim the site as their own. "Are you sure you don't work for the government?" the first man asks again. "You look like a guy from Fish and Wildlife."

The second hut is open on one side where a garage door has been removed, but the walls provide a windbreak. We sweep out musk ox droppings and broken glass before setting up our tent inside. We explore the broken machinery next to our hut. On the ground we find rusted tools, along with scrap metal and wood and tarpaper, all overgrown by tundra grasses. Between the huts and the Cape Thompson bluffs, we discover an old radio shack, its walls collapsed and lying on the ground, its antenna mast sprawled across the tundra, but with the radio itself still standing, as big as a refrigerator, bolted to the building's foundation. The radio's long-failed vacuum tube technology stands exposed to Arctic weather. The size of radios like this one was driven in part by the heat of their components—miniaturized and packed too closely together, they overheated. In the days since Chariot, radio tubes have been replaced by transistors, and transistors have been replaced by silicon chips, but the challenge of dissipating heat remains. Some of today's desktop computers incorporate tiny water-cooling systems to control heat, and engineers are working on methanol-based cooling systems that might soon be used in laptops. For supercomputing centers, with rack after rack of silicon chips working in tandem, air-conditioning failure can spell disaster.

Did Teller use the radio that now lies in the tundra? Did he ride in one of the abandoned Weasels? Did he bathe in Ogotoruk Creek just here where it runs behind the abandoned buildings and out to the Chukchi Sea?

I bathe briskly in Ogotoruk Creek, washing away three days of dirt and sweat and shivering as the wind dries my body.

In August 1959, the Chariot camp had ten dormitories, a mess hall, a shower and laundry with hot water, three generators, a mechanic's shelter, latrines, and two warehouses. It supported sixty-six men, who referred to it as Camp Icy Meadows. The men carried out experiments, including the one in which soil imported from the Sedan shot was buried in the tundra. Sedan soil containing cesium-137, strontium-90, and plutonium-239 was buried in twelve plots within a mile of the Chariot camp to better understand how radioactive remains from the Chariot blast would be likely to move into the creek and through the food chain.

During the half century since Chariot's abandonment, the camp has been used by various cleanup crews. One of the crews sought out the buried radioactive material that had been brought here from the Sedan shot. The government has claimed that the material was removed, that no Sedan radiation remains at the Chariot site. The three Inupiat workers do not believe it. They seem, in general, disgusted by the federal government. They joke that, having worked here all week, they now glow at night. The man who had asked twice before asks again, "Are you sure you don't work for the government?"

I snap a picture of a metal tray lying on the tundra near our shack. It is the sort of cafeteria tray with an indentation for silverware, another for a main course, two more for sides, one for dessert, and one for a carton of milk or a glass of water. Did Teller's men eat from this tray? Or was the tray left by cleanup crews? Through my viewfinder, the tray appears in the foreground surrounded by tundra grass and low-growing dwarf willows, with the main camp in the background. Behind the camp, the sky and the Chukchi Sea stretch out toward Russia.

Waves slap against the gravel beach, their noise drowned by the howling wind.

That evening, at twilight, the moon sits low over the Chukchi Sea. It is a full moon, deep red behind the silhouette of what is left of the Chariot camp. It is the deep red moon that one would expect when the atmosphere is full of the sort of fine particles thrown into the air by distant desert sandstorms or forest fires or volcanic eruptions or nuclear explosions.

In 1959 the government received a letter from the Alaskan Arctic. "We the undersigned," the letter said, "the Point Hope Village Council do not want to see the explosion at the near area of our village Point Hope for any reason at any time."

The government also heard from a resident named Kitty Kinneeveauk: "I'm pretty sure you don't like to see your home blasted by some other people who don't live in your place like we live in Point Hope."

From another Point Hope resident, Joseph Frankson: "You know, anybody that's born anyplace always likes his home. I don't care where people are from. I know I like this place, Point Hope. So I like to keep living here."

After four years of bickering, on August 24, 1962, an Atomic Energy Commission media release announced that Project Chariot would be "held in abeyance." On the same day, the *Anchorage Daily Times* and the *Fairbanks Daily News-Miner* ran an Associated Press article. "Alaskan Eskimos won a victory over atomic science today," the article reported. "Their great white father isn't going to order anytime soon, if ever, a big nuclear boom on their happy hunting grounds." The people of Point Hope would be able to stay in Point Hope. They would not be chased away by radioactivity.

Three years later, in 1965, far out at the end of Alaska's

Aleutian Island chain, more than a thousand miles from Point Hope, the government set up a belowground bomb-testing facility. They detonated an eighty-kiloton hydrogen bomb named Long Shot. It was followed by two more, the one-megaton Milrow shot in 1969 and the five-megaton Cannikin shot in 1971.

Temperatures hotter than the surface of the sun itself momentarily flared up deep under the surface of the windswept Alaskan island known as Amchitka. The little island of Amchitka, once the home of Aleuts, then a military base that survived Japanese bombing during World War II, catalyzed a growing movement to halt the proliferation of nuclear weapons. The Milrow shot inspired the founding of Greenpeace, and in 1971 Greenpeace activists set sail in an old halibut seiner, the *Phyllis Cormack,* to witness and protest the Cannikin shot. Nuclear testing on Amchitka ended that year, and the island was declared a bird sanctuary. Greenpeace went on to advocate against whaling and later launched a climate change campaign.

Back in Anchorage, I telephone the Firewalking Institute of Research and Education, ready now to schedule a walk. But our calendars fail to mesh. I scramble for another opportunity. I find, by luck, a woman in California who plans to lead a walk in the spring. Her fire walks focus on spiritual growth, both hers and her initiates'. Her next fire walk will take place under a full moon.

She considers the quarter cord of firewood typically used on fire walks inadequate. She will burn a full cord of dried and seasoned cedar under a full moon, rain or shine.

"Sign me up," I tell her. But spring is some months away, and I need warmth now. I need some sun.

Chapter 8

THE TOP OF THE
THERMOMETER

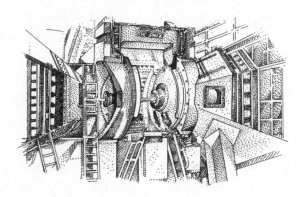

My companion and I fly to Bonaire, near Venezuela, just seven hundred miles from the equator, leaving in our wake several tons of carbon emissions. In three days, Bonaire will become a Dutch municipality. Taxes will go up. Roads might improve.

I sit on a cracked concrete seawall staring out over the reef. A million million thermonuclear explosions warm my face. The explosions occur in the core of the sun, 320,000 miles below its surface, a place where immense pressure and heat push hydrogen atoms together to become helium, a place with a density fourteen times that of lead, a place of Teller's dreams, a fiery furnace broiling at an unimaginable twenty-four million degrees.

Energy emitted in the core moves outward, making its way through a soup of hydrogen, bouncing from atom to atom, transiting toward the surface. The journey might be as quick as fifteen thousand years or as long as a million years. No one knows. What is known is this: the deadly gamma rays and x-rays generated in the core emerge from this soup, by and large, in the form of light and heat. And this: from the sun's surface to my face, the journey requires only eight minutes and sixteen seconds. I sit eight minutes and sixteen seconds from the surface of the sun, but tens of thousands of years from its core.

The sun is a common type of star, its singular claim to importance residing in its proximity to earth. William Herschel, discoverer of infrared light, discoverer of Uranus, wrote of the sun in his 1833 textbook on astronomy: "The sun's rays are the ultimate source of almost every motion which takes place on the surface of the earth. By its heat are produced all winds, and those disturbances in the electric equilibrium of the atmosphere which give rise to the phenomena of terrestrial magnetism. By their vivifying action vegetables are elaborated from inorganic matter, and become, in their turn, the support of animals and of man, and the sources of those great deposits of dynamical efficiency which are laid up for human use in our coal strata. By them the waters of the sea are made to circulate in vapour through the air, and irrigate the land, producing springs and rivers."

Herschel was a brilliant man, but he was not always right. The sun, for example, does not support earthly magnetism. And he occasionally saw heavenly bodies that did not exist. And he believed the sun to be populated. "But now I think myself authorized, upon astronomical principles, to propose the sun as an inhabitable world," he wrote. "It is most probably inhabited, like the rest of the planets, by beings whose organs are adapted to the peculiar circumstances of that vast globe."

A century and a half after Herschel, we know so much more. The sun is not inhabited. Its past and future are well understood.

Thirteen and a half billion years ago—a scant few hundred

thousand years after the birth of the universe, after the Big Bang—the hydrogen that would become the sun drifted in a thin cloud. This cloud drifted through space for nine billion years before it was hit by the shock wave of an exploding star. That shock compressed parts of the cloud, triggering a star-forming contraction. The shock wave also carried with it certain foreign elements—elements like oxygen and carbon and neon, elements that formed in the exploding star, elements that are present in the sun and on the earth. These elements came in proportions suggesting that the exploding star itself had formed from the debris of yet another exploding star. The hydrogen cloud was old, first-generation stuff, stuff from the early universe, but the shock wave that hit it was stuff twice cycled through, entering a new life. Our sun, Herschel's sun, is a third-generation star.

The sun is a star of average size, yet it is difficult to imagine its immensity. In the four and half billion years that have passed since the sun's birth, hydrogen has become helium at a rate of something like four million tons a second.

During these four and a half billion years, the sun has changed. Every billion years, it grows 10 percent brighter. It grows 10 percent hotter.

A billion years from now, the sun's brightness, its warmth, will evaporate the oceans of earth. The oceans, as water vapor, will trap heat. The earth's surface temperature will increase to seven hundred degrees.

Five billion years from now, the sun will exhaust its hydrogen supply. Fusion will slow. With this slowing, the outward explosive forces will be overcome by inward gravitational forces, and the sun's core will contract on itself. With contraction, temperatures will increase. At 180 million degrees—six times hotter than today—the heat will trigger a new round of fusion reactions. These reactions will convert helium to carbon. Slowly, energy from the fusion of helium to carbon will reverse the sun's contraction. The energy will push the sun outward. The expanding sun will swallow Mercury and Venus. The expanding sun will destroy what is left of the earth.

A few more billion years will pass, the sun pulsating, shrinking and expanding. It will throw off its outer skin, forming a sphere of glowing green and red gas. Inside that sphere the remnants of today's sun will collapse on itself, becoming earth sized but more dense. It will become a white dwarf, a crushed ball of carbon and oxygen and heavier elements. Its surface will be pulled smooth by gravity. There will be no hills, no valleys, no craters. It will be a hundred thousand times more dense than steel. A hydrogen bomb, detonated on its surface, would not make a dent.

As a white dwarf, the sun will not be alone. Ninety-seven out of every hundred stars in our galaxy will become white dwarfs, too small to fully explode, too small to become supernovas. Fusion will end. The sun will become a degenerate star, its life defined by trapped heat and by the heat of slow contraction, of gravitational collapse. An inch of contraction provides a hundred thousand years of heat for the white dwarf future sun. But over time it cools. It fades to yellow and then orange and then red and then brown and then, finally, fifteen billion years from now, black.

From the seawall, I launch myself into the water. I swim above coral and fish and eels with the sun on my back. I gawk. The coral here has seen better days. The coral throughout the Caribbean has seen better days. Warmer water and pollution and too many gawkers and maybe hurricanes change the reef. In life, the reef-building corals depend on a partnership with a photosynthetic protozoan, their zooxanthellae. With warmer waters, with less-than-ideal conditions, they expel their zooxanthellae. The coral, without zooxanthellae, becomes white. It dies. Algae grow over its carbonate skeleton. The reef grows patchy.

Thickets of branching elkhorn and staghorn coral that once stood in the shallows are gone, dead and broken up, transmogrified to beach sand. I find a brain coral, boulder sized, its lower half alive and bluish green, its shoulders bleached as white as salt, its cap fuzzy with algae.

Floating facedown, my back and legs and arms expose a combined area of maybe six square feet to the sun. From the sun at this latitude, an area of six square feet receives the energy of ten sixty-watt lightbulbs. I feel the power of ten sixty-watt lightbulbs burning into my skin, warming this water, heating this planet.

The water temperature here is eighty-eight degrees.

Back home, I call my firewalking instructor in California to confirm a time and date. While I have her attention, I ask if she knows anything about William Herschel. Does she know that he believed the sun to be populated? She does not. Does she know anything about extreme heat, about the heat generated by fusion reactions in the sun, by experiments with supercolliders? She does not.

Her interest is in the power of the mind. "It's not enough to hope," she says, "you have to believe." She tells me that her fire walks have led to healings, to the power of the mind overcoming disease. She thinks that attempts to explain firewalking scientifically detract from the experience. Firewalkers are not helped by the knowledge that the coals have little total heat capacity and therefore little heat to release, nor are they helped by the knowledge that the ash layer surrounding hot coals offers protective insulation. They are helped by the power of the mind. They are helped by belief in their own abilities.

I ask her to recommend a book about firewalking. She discourages me from reading ahead of time. She does not want me to develop preconceived notions.

I hang up the phone and read Loring Danforth's *Firewalking and Religious Healing*, mainly about the Anastenaria of Greece, where the healing effects of firewalking are well known, but also about the American firewalking movement. Danforth quotes one firewalking instructor, an American: "I've led over three thousand people through fire, and I've had two people in bed for a week with third-degree burns." A seventy-year-old woman walked through fire and was hospitalized with second-degree burns. The president of a major corporation walked through fire and was burned. "I got burned," he later said. "And so what? It became a positive experience."

The book includes a copied advertisement for one of Tolly Burkan's fire walks, with a parenthetical notice: "Professional fire fighters admitted free with I.D.!" It also includes a quote from a man who had just walked through fire: "I am convinced that if you had your shit together and really put your mind to it, you could survive a direct nuclear blast."

I fly to Rome on business. It is raining, the streets full of Romans with umbrellas fully deployed. They talk of a new record low temperature for Italy, a temperature achieved on the day of my arrival, three hundred miles north of Rome in the Italian Alps, the Dolomite Range, in a low spot, in a frost hollow. The temperature there touched fifty-four degrees below zero.

In passing, I talk to a physicist at the Sapienza University of Rome, founded in 1303 by Pope Boniface VII. *Sapienza:* knowledge or wisdom, so it is the University of Knowledge, just as *Homo sapiens* is the wise man.

Enrico Fermi was among those who found knowledge here. With this knowledge he fled fascism, accepting a Nobel Prize in Stockholm en route to Columbia University, where he joined Leo Szilard. Szilard was the man who, with Edward Teller, would convince Einstein to warn President Roosevelt of the possibility of nuclear weapons, triggering the sequence of events that would lead to the Manhattan Project and the nuclear bomb.

At Columbia, well before the bomb became a reality, Fermi and Szilard proved the possibility of nuclear chain reactions. They proved that a neutron bombarding a uranium nucleus could split the uranium atom, sending out more neutrons, which could in turn split more uranium atoms. They moved to the University of Chicago, where they built the world's first nuclear reactor in the racquetball courts beneath the football stadium. It was a model for the secret reactor that would be built in Hanford, Washington, the reactor that would manufacture the plutonium used in the Fat Man bomb dropped on Nagasaki. With Robert Oppenheimer, Fermi is sometimes referred to as the father of the atomic bomb, just as Teller is referred to as the father of the hydrogen bomb.

Fermi knew Teller well. They played Ping-Pong together. It may have been Fermi who suggested to Teller that a fission bomb could be used to trigger a fusion reaction, that an atomic bomb could make a hydrogen bomb possible. Fermi was also among the physicists who worried that nuclear explosions could ignite the atmosphere.

But the Italian physicist I talk to now is no Fermi. He does not believe that the world's climate is warming. As evidence against it, he cites the record cold temperature from the Dolomite Range. It is not the argument of an educated man. He ignores John Tyndall's nineteenth-century measurements showing that carbon dioxide traps heat. He ignores five decades of data showing increasing temperatures. He ignores climate itself, focusing instead on the vagaries of local weather.

I break away. I wander alone on the sidewalks of Rome in the rain. I walk past the Coliseum, completed the year after Vesuvius exploded, the year after Pliny the Elder died. I wander through the ruins of the Forum, imagining the stout Pliny, attended by servants and admirers. My feet touch ground on which Pliny once trod. Even then, two thousand years ago, he would have seen ruins here, temples built seven hundred years before he was born, buildings that had been sacked and rebuilt.

I walk fifty minutes through the rain to Saint Peter's Square. I walk past the obelisk in the courtyard, navigating around pigeons and tourists, stepping on ancient cobbles and soggy cigarette butts. I find a docent who speaks English. I ask if she can direct me to the offices of the Congregation for the Doctrine of Faith. She does not know where their offices are but she asks another docent, who asks another, who asks another. Within seconds four docents stand in a circle. A complicated and heated discussion ensues, but after five or six minutes the woman returns her attention to me. She smiles. She says they are not sure, but they think it is the building just there, "down and to the right, past the Swiss Guards." She gestures, a fingernail darkly painted pointing the way to a nondescript building, five or six stories tall, with shuttered windows, the offices of the Congregation for the Doctrine of Faith, the Congregatio pro Doctrina Fidei.

I talk to the Swiss Guards themselves, young men, alert, wearing sharp black berets but otherwise dressed like clowns, their official uniforms striped in red and blue and orange, topped with wide white collars. They want to be helpful, but the building is off-limits unless I have official business there. I do not, so I sit for a few minutes on the concrete base of a column, staring at the building.

In 1633, Galileo Galilei was tried by the Congregation for the Doctrine of Faith, known then as the Supreme Sacred Congregation for the Roman and Universal Inquisition. His alleged crime: "holding as true the false doctrine taught by some that the sun is the center of the world." More precisely, Galileo was tried for ignoring an earlier injunction by the church, an injunction ordering him to abandon heliocentrism, to give up blasphemous thoughts that would inconvenience the church.

The building in front of me is not old enough to have been the site of Galileo's trial. But somewhere near here, under threat of torture, on June 22, 1633, Galileo spoke: "I, Galileo, son of the late Vincenzo Galilei, Florentine, aged seventy years, arraigned personally before this tribunal, and kneeling before you . . . swear that I abandon the false opinion that the Sun is the center of the world and immovable, and that the earth is not the center of the world, and moves."

He was sentenced to imprisonment, but the sentence was relaxed to one of house arrest. He died nine years later, still under house arrest.

A century passed. In 1758 the church dropped its ban on writings claiming that the earth circled the sun. Two more centuries passed. In 1992 the pope called Galileo's conviction "a tragic mistake."

For a time the Vatican considered erecting a statue of Galileo. From Nicola Cabibbo, a nuclear physicist and head of the Pontifical Academy of Sciences, in 2008: "The Church wants to close the Galileo affair and reach a definitive understanding not only of his great legacy but also of the relationship between science and faith." But the statue would not come to pass. The church diverted the statue's funding to Nigeria, to help Nigerians better understand the relationship between religion and science.

I try to imagine the great man himself, a septuagenarian, his weathered face strong behind a gray beard, worried about his immediate future, his fate to be dictated by the politics of

Catholicism, by the Inquisition. Even then, it is easy to imagine him obsessively thinking about the planets and stars.

A nun walks past carrying a shopping bag, and the Swiss Guards step aside. Two more nuns drive up, and the guards wave them through. In front of the building itself, the building that houses the offices of the Congregation for the Doctrine of Faith, a policeman dressed in dark blue and wearing a thick Kevlar vest carries a machine gun. A middle-aged woman panhandles. Church bells ring.

Back home, I place another telephone call, this one to a physicist known for her work at the Brookhaven supercollider, officially called the Relativistic Heavy Ion Collider. She is a distinguished professor, known for smashing the nuclei of gold atoms together at close to the speed of light just outside New York City.

The goal, she says, is to learn more about quarks, to see quarks in isolation. Quarks are strange things, otherworldly, behaving in ways that make no sense at all in the everyday world. They are the stuff of protons and neutrons, building blocks that do not, under anything resembling normal conditions, exist in isolation. In day-to-day life, they exist only in groups. The groups are called hadrons, and protons and neutrons are two kinds of hadrons.

"In the normal course of events," the physicist tells me, "heat transforms matter from solids to liquids to gases." Heat moves the molecules farther and farther apart, forcing them to dance. Add more heat, and electrons move away, leaving a plasma, a soup of positively charged ions and electrons. In plasmas, matter remains intact. The nuclei of atoms remain unchanged. The protons and neutrons retain their integrity. Plasmas are not especially exotic. Lightning is a plasma. The imprudent homeowner can create a plasma in a household microwave oven with nothing more than a match, a bit of electricity, and acceptance of the possibility of a fire. Plasmas are normal, just another phase of matter.

But add enough heat, and another phase change occurs. Protons and neutrons fall apart, breaking into their constituent parts. Quarks are freed.

"The early universe," the physicist says, "was all about quarks."

The early universe—the first microsecond, the first flash—was unimaginably hot. The heat just after the Big Bang did not allow the existence of normal matter.

But the early universe is gone. It is fifteen billion years in the past. "It is not something you can see," she tells me. "It is not something that astronomers can study."

And so people like her use supercolliders to smash together heavy ions, re-creating the temperatures of the early universe. She casually mentions that the experiments at Brookhaven recently reached a temperature of four trillion degrees Kelvin, or more than seven trillion degrees Fahrenheit. This is the highest temperature ever created in a laboratory. It is hotter than the core of the sun. It is hotter than the core of a supernova, which itself is thousands of times hotter than the core of the sun.

I make plans to visit Brookhaven, the last stop on a journey to the top of the thermometer. But first there is the matter of firewalking.

With my companion, I fly to California. With two days between us and our firewalking appointment, we decide to walk in the mountains. We pick an area near Nevada City, a Gold Rush town once populated by forty-niners. Some of the Death Valley survivors may have found their way here. They may have bought a bottle of Kier's Petroleum. Some of them may have seen Mark

Twain here when he spoke in 1866, describing his travels in Hawaii, describing his visit to Kilauea.

We drive on gravel roads wet with steady rain. The mountains have had their faces ripped off, torn away by hydraulic mining to leave behind steep barren bluffs. The forty-niners created landscapes as barren as those of Death Valley, or more so, but wetter. We park our rental car and walk the Humbug Trail, named by forty-niners who failed to find gold. The rain turns to snow— fat flakes that settle lazily to the ground. When we return to our rented car, the flakes lie stacked four inches thick on the road. The car refuses to climb hills through the snow. A man and his wife offer us a ride. We abandon the rented car and return to Nevada City to wait for our fire walk while nature mocks us with ever-thickening snow, throwing down barriers between us and hot coals.

But we are not to be stopped. We find alternative transport to the Goddess Temple, an hour from Nevada City and below the snow. The temple stands on private land at the end of a long driveway. To enter the temple, one walks beneath the open legs of a two-story metal woman, a sculpted New Age goddess. The goddess stands soaking wet, pelted by rain, blasted by wind. Behind her, in an open space a hundred feet from the temple itself, blue tarps cover a full cord of wood. Rain has pooled on top of the tarps. Wind whips the tarps, slapping them against the wood.

Inside the temple, the rain finds its way through a leak in the ceiling to a pan on the floor. A woodstove in the corner, in the shape of an elephant and as big as an elephant calf, heats the room. Firewalking novitiates mill about, waiting, hovering near the stove for warmth. A stereo tuned to a top-forty station competes with the sound of wind at the windows.

I find my instructor, and we talk quietly. She reminds me that she burns a full cord of wood for firewalks. "I want firewalkers to have the full experience," she says.

And this: "My goal is to inspire personal empowerment without getting burned."

Getting burned remains a distinct possibility. The firewalkers, me included, sign waivers. The waivers, the instructor tells me, hold up in court. "The courts know that anyone walking through a fire is taking on some risk. It's like skydiving."

Some firewalking events, she says, cater to a thousand participants. These are the corporate team builders, big events set up for big businesses. Tonight she has thirty-four walkers signed up. This is not a team builder. This is about personal growth. This is about healing.

"No one has to walk through the fire," she tells me. "There is no pressure. Some people will just watch, but even in watching they will feel empowered. The fire's presence in itself is inspiring."

But she is there to help people break through their personal barriers. She once told a teenage girl, "You don't have to do anything you don't want to do." The girl did not want to walk through the fire alone, but she wanted to walk through the fire. The instructor held her hand. They walked through together.

We talk for a time about the science of firewalking, about physical explanations of firewalking. Hot coals conduct heat poorly. The feet are in contact with the heat for fractions of a second. Walkers are unscathed because of physics, not because of a mystic force. But in the end, it is about a barefoot walker and a bed of glowing red coals. It is about the very real possibility of being burned. It is about intense heat and the human reaction to the presence of that heat.

"The science doesn't really matter," she says. "In some ways the fire itself doesn't matter. It's not about the fire. It's about conquering fears. It's about healing. It's about overcoming personal barriers."

When she is not guiding people through fires, she is a clinical behavioral therapist. She has helped people with depression, with anxiety, with panic attacks, with sleep disorders, with addictions. She has counseled cancer patients. She has counseled victims of rape and child molestation. What she does here, with firewalking, she sees as an extension of what she does there, with behavioral therapy. She helps people help themselves.

Thirty-four of us sit on cushions on the floor of the Goddess Temple. The instructor, up front, talks about firewalking. She talks about fears and limitations. She talks about Saint Francis of Paola, also known as Saint Francis the Firehandler, who held burning coals. She tells us that the Romans, in the time of Pliny the Elder, did not tax those who could walk through fire unscathed. Now firewalking has come back. People are tired of being told that happiness comes with the purchase of a particular car, with the wearing of a particular pair of jeans, with the eating of a particular cereal.

She talks of a certain tribe in which men not only walk through fire but roll in it, rubbing the hot coals against their own bodies and those of their fellow tribesmen. "We won't do that," she says. "All we are going to do is walk through the fire."

As a group, we move outside. For the moment, the rain has slowed. We watch as the blue tarp is moved away from the wood. The instructor's assistants spray lighter fluid across the top of the pile and along its edges. A flare is ignited and in turn is used to touch off the lighter fluid and the wood. Within minutes, a raging bonfire dances in the wind.

While the fire burns, we return to the temple. The instructor offers further inspirational comments. She talks of walking on recently hardened lava. She talks of building the energy within yourself that will make this fire walk possible. While she talks, one woman, sitting on a cushion, knits. A man, sitting on another cushion, meditates. Two women, on separate cushions, hold hands.

The instructor quotes Henry Ford: "Whether you think that you can, or that you can't, you are usually right."

And she tells us that there will be no layer of ash between those hot coals and our bare feet. "Despite this," she says, "you are not going to feel the heat on the soles of your feet. You might feel the heat rising up from the bed of coals on your legs, and maybe on your faces, but to your feet it will feel like you are walking on popcorn. There will be no pain. It will just feel a little crunchy."

She demonstrates the technique. "Don't run," she says. "Don't stamp. Don't dance. Just walk across." Near the front of the temple, she has each of us walk across an imaginary fire, one at a time.

And then we are back outside, standing in a half circle around the fire. The rain has slowed to a drizzle. From a distance of five feet, the heat of the fire offers comfort, but the gusting wind renders the flames unpredictable, sending out unexpected flares. My feet, bare, grow cold in the mud.

The instructor's assistants rake hot coals on the ground. They build a thick layer of glowing embers and burning wood. They make an incandescent promenade. Along this promenade, a flame pops up and goes out, and then another, and another. This is the real deal.

It is time. I am the second in line waiting to walk across hot coals. I am the second in line waiting to become a firewalker.

In an ice-covered tent, in spruce forest or on frozen tundra or on the sea ice, heat becomes a central focus, like hunger for a starving man or water for Pablo Valencia. It is impossible to think of anything but heat. The same is true now, but for very different reasons. Here at the edge of the fire pit, my mind is supremely focused. For a moment, my mental space, my conscious self, is consumed by fire and heat. I understand the science. I understand that hundreds of thousands of people have walked across hot coals unscathed. But I also understand that I am about to step barefoot onto a bed of glowing coals and flaming wood with a temperature of one thousand degrees. I understand that what I am about to do feels just the slightest bit insane.

And so I step onto the bed of coals and cross fifteen feet of thousand-degree ground, barefoot. The heat rushes up my pants legs. I feel it on my hands. I feel it on my face. The burning ground feels rough, like rounded gravel, like popcorn, but it does not hurt my feet. A moment later I do it again, and again, this time hand in hand with my companion. On the third walk, I step on something hot—a sharp ember or a burning stick—but I keep walking. I smile broadly. I am a firewalker.

Later I talk to the instructor about my feelings as I walked across the fire for the first time. I was surprised by my hesitancy, but I was also surprised by my sense of exhilaration.

"Your mind has to be in a certain place," she says.

And I jot down this simple note: "In firewalking as in life, your mind has to be in a certain place."

The physicist, in another telephone conversation, mentions that there have been problems with litigation. The road to the top of the thermometer is not without legal hurdles.

The plaintiffs presented a straightforward premise. Collider experiments explore unknown ground in physics. It is possible, the plaintiffs argued, for the experiments to create runaway reactions capable of massive destruction. For example, tiny black holes could form, and over time they could consume the earth. Alternatively, strangelets could materialize. Strangelets are a hypothetical form of matter that could convert all other matter into something similar to itself, again consuming the earth.

The plaintiffs invoked the National Environmental Policy Act of 1969, signed into law by Richard Nixon in the wake of the Santa Barbara Channel oil spill.

"There is no question," the plaintiffs explained to the court, "that should defendants inadvertently create a dangerous form of matter such as a micro black hole or a strangelet, or otherwise create unsafe conditions of physics, then the environmental impact would be both local and national in scope, and deadly to everyone."

The plaintiffs demanded an environmental impact assessment. If someone was going to play around with the fundamental building blocks of the universe, if someone was going to cook quark soup, the plaintiffs wanted to understand the potential environmental impacts.

Some critics said that at least one affidavit provided to the court appeared to be based on a Wikipedia article.

Many members of the experimental physics community considered the plaintiffs to be crackpots. The kinds of collisions created in colliders happen with some regularity in nature, when cosmic rays smash into particles in the upper atmosphere, with neither local nor national environmental impact. The earth has not been devoured.

The difference between the collisions that occurred in the upper atmosphere and those that would occur in the collider was one of instrumentation. To understand what happens during high-speed collisions, physicists need to observe thousands of collisions in one place, and they need instruments that weigh four thousand tons. The upper atmosphere, with its collisions spread out in space and time, without a firm foundation for instrumentation, was no place for measurements.

The court closed the case on various legal grounds. The experiments moved forward.

❈

Edward Teller, when he was not making bombs, occasionally thought about climate change. In his memoir he wrote of a conversation with Dixy Lee Ray. Ray was a marine biologist and an environmentalist, but also an advocate of nuclear power. Under Richard Nixon, she served as chair of the Atomic Energy Commission. Later, when she was governor of Washington, Mount St. Helens erupted.

From Teller's memoir: "Dixy and I touched on several topics over lunch that day, one of them the question of whether the increased level of carbon dioxide in the atmosphere could cause significant global warming."

From another passage in his memoir, in which he advocates nuclear energy: "Meeting global energy needs by an increased use of fossil fuels will not only increase atmospheric carbon dioxide (which may be involved in causing climate changes) but also may lead to the exhaustion of an economic coal supply."

Teller and Ray, if they were still alive, would be sorely disappointed by the tsunami that damaged reactors in Japan. They would see the tsunami as a dreadful setback, as an unfortunate event giving nuclear power yet another undeserved black eye. Even mainstream environmentalists, they might think, were beginning to see the light, and now this wave has thrown us back onto the beaches of a carbon-dependent world.

❈

I pack my bags one more time, off for a look inside the supercollider at Brookhaven. I want to see the place where humans have created and controlled temperatures of seven trillion degrees.

The journey involves an airplane ride to the East Coast, with more burning of kerosene and more carbon released, followed by a train ride to Long Island. I ask a conductor if he has ever heard of Colonel Edwin Drake, the driller of the Titusville oil well, the man credited with starting the petroleum economy. The conductor has not heard of Drake.

"He was a conductor, too," I say, "before he went into the oil business." The conductor looks at me for a second, says nothing, and moves on, punching tickets.

At the railway station I meet a courtesy van, and thirty minutes later I am in the Brookhaven complex. What is now the Brookhaven National Laboratory once belonged to the U.S. Army. It was Camp Upton, the onetime home of the World War I draftee Irving Berlin and the inspiration for his "Yip, Yip Yaphank" and "Oh, How I Hate to Get Up in the Morning."

The base was deactivated after the war to end all wars and reactivated for the next war.

Soon after the Hiroshima bombing, scientists from Yale and Harvard and Cornell and six other universities pushed for a new laboratory, a place that could house big experiments, a home for jointly purchasing and maintaining equipment that no single university could afford on its own. They wanted a place for big science. They found Camp Upton in the backwoods of Long Island.

In March 1947 the War Department handed the camp over to the Department of Energy, and the Department of Energy turned the scientists loose. It would be six years before Eisenhower would deliver his Atoms for Peace speech to the United Nations, but from the beginning the Department of Energy saw Brookhaven as a place for peacetime science, for investigations of the atom that would contribute to the well-being of society. The Department of Energy beat the swords of Camp Upton into plowshares.

Now students and professors come and go, some for a day, some for a few months, some for years. Brookhaven has become an IQ magnet, where smart people congregate to work on things that excite geniuses. Work on the atom continues, but also climate research and medical research and energy research. They are building a solar farm to provide both power and knowledge. They work across disciplines. Biologists work with engineers who work with nuclear theorists who work with computer scientists who work with chemists.

In a Brookhaven building, I ask three men to tell me what they do. Are they computer scientists? Are they physicists? They hesitate. They are not sure. At Brookhaven, highly educated specialists are no longer certain of the boundaries of their knowledge.

And there is the supercollider.

From the outside, it looks like a river levee, a sloping mound of dirt covered with patchy grass and weeds and shrubs. But there is no river. The levee covers the tunnel that is the heart of the collider. The tunnel forms a circle on Brookhaven's grounds. The circumference of that circle, around which protons and the nuclei of gold molecules fly, is about 12,600 feet, slightly shorter than the Indianapolis Motor Speedway.

On the Indianapolis Motor Speedway, drivers sometimes push their cars to speeds of 235 miles per hour. In Brookhaven's tunnel, physicists sometimes push their gold nuclei to speeds of 670 million miles per hour.

At Indianapolis, as a general matter, the goal is to avoid collisions. At Brookhaven, as a general matter, the goal is to encourage collisions.

At Indianapolis the traffic flows in one direction. At Brookhaven traffic flows two ways, clockwise and counterclockwise, in two separate beams. The physicists, with magnets, occasionally make the beams cross. That dangerous intersection, with neither traffic signals nor stop signs, is where the action is.

꙳

Building 1005 stands above the tunnel. I meet the physicist in front of Building 1005. I worry that I am wasting her time, a tourist of heat interfering with high-energy physics, but she smiles. She, too, enjoys visiting the collider. Although she has worked on it since its beginning, even during its design, there are parts of it that she has not seen. This is the nature of big science. It involves brigades of scientists and engineers, all working together, some never meeting in person but melding the output of their brains to create a collective capable of something like this, capable of seven trillion degrees. She is as interested as I am.

And when her experiments are running, she cannot visit the tunnel. No one can. When the experiments are running, when gold nuclei race around the circuit, occasionally slamming into one another and disintegrating into their elemental early universe stuff, the tunnel is completely off-limits.

Before we go to the tunnel, to the hottest of all places, we tour the refrigeration plant. To make the hottest of all temperatures requires something approaching the coldest of all temperatures. The particles that these geniuses collide have to be accelerated, and that acceleration uses powerful magnets—1,240 of them—and those magnets require the superconductivity that comes with temperatures approaching absolute zero, temperatures approaching 460 degrees below zero Fahrenheit.

We look over the banks of compressors and heat exchangers. Today they sit idle, quiet. During experiments, in operation, they create a deafening din. In principle, it is the same din that comes from a home refrigerator, but scaled up. And instead of Freon or some similar refrigerant, it uses liquid helium, and instead of striving for temperatures at which one can safely store ice cream and salmon, it strives for the very cold temperatures needed to promote superconductivity.

The walk from the refrigeration plant to the tunnel requires no more than five minutes. We go down steps and around a corner into the tunnel itself. Radiation that escapes from the particle beams will travel in a straight line, hitting a concrete wall, leaving anyone around the corner secure.

There are measures to ensure that no one is inside during experiments. There is a badge system, a security camera monitors the coming and going of workers, and access depends on an elaborate key system. No one starts this car unless everyone is outside the tunnel.

The inside of the tunnel itself resembles an underutilized utilidor, big enough for a single lane of traffic, but it holds nothing more than two pipes and a few wires and a rack hanging down from the ceiling, above the pipes. The entire tunnel is scrupulously clean. One could, without hesitation, eat from its floors. The tunnel's walls curve like the gentle curve of a railroad track. The pipes, at this location, run side by side as far as I can see along the curving route. But elsewhere, beyond my line of sight, they cross.

The pipes, on stands that hold them waist high, are eighteen inches in diameter. A blue racing stripe marks the pipe on the inside track, indicating that its particles will travel clockwise. A yellow racing stripe marks the pipe on the outside track, indicating that its particles will travel counterclockwise.

Each pipe, for the most part, holds insulation. Inside that insulation, smaller pipes hold the liquid helium that cools the magnets. Those magnets surround another pipe, an innermost pipe. This innermost pipe, a few inches in diameter, runs through the center of the outer pipe.

Inside the innermost pipe, for the most part, there is nothing. It holds a vacuum. But in the center of that vacuum, in the center of the innermost pipe, itself in the center of the outer pipe, a beam with a diameter of four one-thousandths of an inch occasionally flows. That beam is occasionally full of flying particles. The particles are gold.

The physicists convert elemental gold into gold ions, gold molecules stripped of their electrons. Those gold ions are sent into that tiny beam in tiny batches. Inside the beams that are inside these pipes—the beam in one pipe going clockwise, the beam in the other pipe going counterclockwise—batches of gold ions are pushed toward the speed of light.

Where those pipes cross, where batches of ions meet in the most dangerous of intersections, gold ions are annihilated. This is not merely a matter of fission. The ions are not merely turned into smaller atoms. They are not merely turned into scattered protons and neutrons. Parts of them are annihilated, some of their protons and neutrons converted to quarks, the particles that make up protons and neutrons. Protons and neutrons are destroyed.

In the lifetime of this supercollider, the physicists will go to great lengths to destroy a few grams of gold. Collectively, the gold they destroy would not be enough for a wedding band. But somewhere out of sight, down this tunnel, around the curve, where the gold becomes quarks, physicists have observed temperatures of seven trillion degrees.

We drive to the PHENIX detector. It sits in a room the size of a gymnasium, but it is a gymnasium with two stories of hardware, all with one purpose: to intercept the debris cast out by gold ions colliding at close to the speed of light, to collect data, to understand.

The hardware looks something like the armature of a giant electric motor. Or like a hospital's CAT scanner. Or like a metal press of the kind used to crush scrap metal. Or like all of these and none of them. The PHENIX detector is in fact unique, custom built from the ground up at a cost of something like a hundred million dollars, with its metal surfaces painted green and yellow and blue or buffed to an unpainted shine, with cables of black and red.

Now, between experiments, the detector is open, its guts exposed for maintenance. In the middle of it, surrounded by machinery, a single pipe stands out, a tube a few inches in diameter, well above my head. When the experiment runs, monumental collisions occur inside the tiny beam inside that pipe. They have occurred not once or twice but millions of times, at rates of tens of thousands per second. Most are glancing blows, ionic fender benders. But some are head-on, full speed, front bumper into front bumper, headlights into headlights, no braking, no swerving, a game of chicken played right through to its shattering climax.

I stare at the point of collision, at the tube that, when the experiments are running, holds the action. I am distracted by it. Its presence converts me into a rubbernecker. But it is more than that. I am more like a novitiate drawn to the altar. For a moment, I cannot take my eyes away from the tube. In looking, I trip over a rail on the floor and almost fall. I embarrass myself by staring at the tube, drawn in by what has happened here. In this tube, a phase transition occurred, similar to the phase transition from ice to water, from water to steam, but here it was from matter made of protons and neutrons to something else, something different from normal matter. It was a phase transition that left behind a quark soup at seven trillion degrees.

Yet there is nothing to see, nothing but a shining pipe sitting idle and surrounded by machinery. Each collision lasted no longer than a fraction of a fraction of a fraction of a second, and although gold nuclei are large compared to the nuclei of more common elements, they are small. The heat dissipates long before it reaches the walls of the tube, like the heat from a burning race car dissipating before it reaches fans in the upper bleachers. The tube that has contained seven trillion degrees stands intact, inanimate, unassuming and inexpressive of what went on within its walls.

The particles of interest to the physicist, the debris from head-on collisions, fly through the walls of the pipe. The debris moves through the spaces between the molecules of metal just as light moves through glass. Some of them—enough of them— are intercepted by detectors. Those detectors send their signals to computers. Those computers generate beautiful starburst diagrams that show the track lines of debris. And from those track lines, physicists know, among other things, the temperature at the point of collision. It is as though they have an infrared thermometer capable of reading a temperature from the tiniest of points, for the tiniest fraction of a second.

The strange thing, the physicist tells me, the really weird thing, is what they found at seven trillion degrees. They found something unexpected, something that the theorists had not predicted. They found that materials at this temperature behave as almost perfect liquids, as strongly coupled plasmas. They found a quark soup that flowed without resistance, with almost no viscosity.

Liquids like this have been found before, but never at extremely high temperatures. Liquids like this, nearly perfect liquids, liquids with almost no viscosity, are known to form at temperatures approaching absolute zero.

The physicist and her colleagues must have felt much the same as the physicist Isidor Rabi did in the 1930s. Confronted with the unexpected discovery of subatomic particles called muons—particles that were not predicted by then current theory, particles that did not fit with then current theory—Rabi made a remark remembered by every particle physicist. His remark: "Who ordered that?"

Today string theorists work to understand what the behavior of quark soup means. They work with equations that make no sense, forcing the math to explain what remains, for now, unexplainable.

"What comes next?" I ask the physicist. And she understands my meaning. She knows that I am asking if there will be higher temperatures, and what these higher temperatures might do. She says that the collider in Switzerland, the Large Hadron Collider, has a circumference eight times greater than that of the Brookhaven collider. It smashes lead ions into one another with fourteen times the power of the Brookhaven collider. Nothing has been published yet, nothing is for sure, but she believes the physicists at the Large Hadron Collider may already have achieved temperatures 30 percent higher than hers.

"Quarks are thought to be the fundamental building blocks of the universe," she tells me. But her voice carries a thread of doubt. It may be that quarks are the end of the road, that no amount of heat will reduce quarks to something smaller, to something even more elusive. On the other hand, every time that physicists have found the smallest building blocks, something smaller has come up. There were molecules, and then atoms, and then protons and neutrons, and then quarks. And now what?

Two centuries ago, Antoine Lavoisier thought of heat as a subtle fluid. A century ago, the theoretical groundwork behind the atomic bomb was framed. A half century ago, the first nuclear bomb was exploded, and seven years later the first hydrogen bomb. Now controlled collisions annihilate protons and neutrons and generate seven trillion degrees of heat in the name of peacetime research. And a hundred years from now?

Tyndall, in his biography of Faraday, felt compelled to justify science. He talked of electric wires crisscrossing London streets. "It is Faraday's currents," he wrote, "that speed from place to place through these wires." But in contrast, this: "What has been the practical use of the labours of Faraday? But I would again emphatically say that his work needs no such justification and that if he had allowed his vision to be disturbed by considerations regarding the practical use of his discoveries, those discoveries would never have been made by him."

As for myself, I play with candles and walk through fires and read the words of long dead writers, I burn pilfered peat, I stare at Bronze Age corpses, I drink oil, and I am forever enthralled by Brookhaven's pipes. At the top of the thermometer, beyond any temperature that I could possibly imagine, those pipes explore conditions near the beginning of the universe. And across the water, in Switzerland, a larger collider sends lead nuclei hurtling toward one another—lead instead of gold, so heavier particles, traveling just as fast or faster, smashing into one another in ways that will create still hotter temperatures, reaching back in time just an instant closer to the birth of the universe, to a time that remains entirely mysterious and speculative. In my day-to-day life, bundled in a thick coat or standing before my woodstove or moving along a snow-covered trail, I find myself thinking of those pipes. And when I think of them, I remember that at the top of the thermometer lies matter with the audacity to behave as though it were absolutely cold, flowing like a perfect liquid, and without a doubt doing so in a manner that will change the way in which people understand the history of the early universe, that time before the first second passed, before protons and neutrons, before hydrogen and stars, before our earth warmed and cooled and warmed again.

I think of those pipes and their innermost beams and all that has come before them, and I try to imagine what might come next, where our understanding might take us, only to remember that no one knows, that none of us can imagine what comes next, that the brightest of physicists were surprised and puzzled by what they found at seven trillion degrees. I think of those pipes and their innermost beams, and what came before them, and what might come next, and a sense of wonder overwhelms me. Those pipes and their innermost beams leave me basking in awe.

ACKNOWLEDGMENTS

How can I say thank you to people who have changed the way I think about the world, the way I see my surroundings, and the way I experience day-to-day life? For that is exactly what happened when I wrote *Heat*. With each page, with each paragraph, and often even with each sentence, I learned to see heat everywhere I looked in ways that I had not previously considered.

Most of the people I talked to in the course of researching *Heat* have no way to know how they impacted my life. To them, I was a writer with whom they spoke for a few hours or a day or a few days. I was a voice on a telephone. I was an e-mail address. And that is often how life works: the greatest impacts result from the smallest deeds.

I contacted Betsy Gallery, a California-based artist, when I heard that she was working on a fire mural. She was one of the first people I interviewed about heat. At that time I did not have a clear idea of how the book would take shape or what might be included or excluded. Despite this, and although she had no way to know who I was, she was immediately helpful and seemed to quickly understand my interest in heat. As I worked on the book, she completed her fire mural, which now hangs at the Santa Barbara County fire headquarters. Although we did not talk frequently after our face-to-face interview in California,

I often thought of her working long hours in her studio, converting fire artifacts into fire art, as I worked long hours at my desk, converting words about heat into sentences about heat. The knowledge of our somewhat parallel efforts occasionally offered me solace during the inevitable rough periods that come with writing a book. Also, her explanation of the process of designing and building a mosaic from fire artifacts, in which the placement of any one piece would affect all the other pieces, in which thinking ahead was important but would not replace repeated rearrangements, helped me think about the way in which I arranged and rearranged *Heat*.

Betsy Gallery introduced me to her son, the photographer David Gala, whose house, along with most of his photographs, had burned in a chaparral fire. I was reluctant at first to talk to David about heat. It seemed to me an act of intrusion—an act in which I would gain something from his loss. But he did not see it that way. He drove with me to the remains of his home. He introduced me to neighbors who had also lost homes. He showed me how he shot pictures of fire artifacts, and he showed me a collection of photographs that had escaped the fire, having been stored elsewhere when the house burned. And he was a gracious host to a traveling writer whom he barely knew.

Professor Dar Roberts of the University of California in Santa Barbara took time away from his research and family to teach me about burned landscapes and landscapes that have not yet burned—seemingly the only kind of landscapes in southern California. Despite the financial disarray of the University of California system at the time of my visit, and the resulting uncertainty faced by any academic working in that system, Dar was generous with his time and very patient with what must have seemed to be naive questions.

The firefighter Louis Del La Rosa spent a long day with me in the field, pointing out features that only a seasoned firefighter would see. At Spanish Ranch he guided me across ground that he has toured more than once, patiently explaining what happened there during the fire that ultimately killed four men. He humored my wish to test a fire shelter tent in a parking lot even though I am sure that it resulted in a few raised eyebrows from his colleagues. Before I met Louis, I had great respect for firefighters, and that respect grew with every hour I spent talking to him.

My two firewalking instructors, Matt Carr and Claudia Weber, did not know each other, but they shared their enthusiasm for fire and, by extension, for *Heat*. I did not meet Matt Carr in person but hope to do so in the future, perhaps over hot coals. Nevertheless his willingness to spend hours in conversation over the telephone contributed to my understanding of the firewalking movement, the act of firewalking itself, and what I came to think of as the culture of firewalking. The culture of firewalking can be described in part as one of enthusiasm. Firewalkers are enthusiastic about firewalking, but this enthusiasm spills over into all other aspects of life as well. And it is not a selfish enthusiasm but an enthusiasm for all people and for the power of humans to realize that they are capable of more than they might believe. I saw this same enthusiasm in Claudia, whom I did have the pleasure of meeting when she led me through my first fire walk. Claudia was generous of her time both over the telephone and in person, despite the distraction of managing a group of thirty-four firewalking students in the midst of a pouring rain. Her merging of spirituality with firewalking inspired my own thinking about firewalking and allowed me to see past my training as a scientist to enjoy the experience of firewalking as a human being. I was also quite moved by the obvious impact her efforts had on many of the firewalkers present the night of our walk in central California.

Matt Patrick of the Hawaiian Volcano Observatory corrected

misconceptions that I had about the workings of volcanoes. Like many young scientists, he was clearly overworked but was still happy to take time away from his computer and his instruments to talk about volcanoes. He telephoned one evening, after we had been talking most of the afternoon, concerned that I might slip past the warning signs to get a close-up view of the Kilauea caldera. I found his concern for my safety endearing, and in fact when he telephoned he interrupted a discussion I was having with my companion in which we were weighing the pros and cons of exactly the thing he had called to warn us against. Importantly, he also taught me to walk across newly formed and still-hot ground, something I would never have done without his lead, but something I still value as one of the finest of many fine experiences I enjoyed while researching *Heat*.

Vern Miller and Kip Knutson hosted my visit to the Tesoro oil refinery in Nikiski. Without Tesoro's hospitality, I could not have presented a firsthand explanation of the refining process. Vern's knowledge of refining and his enthusiasm for the topic were impressive. Kip's view regarding the importance of transparency was refreshing and not at all in keeping with the reputation of the oil and gas industry.

The physicist Barbara Jacek, with the help of Karen McNulty Walsh from the Brookhaven Media and Communications Office, understood why I needed to see the inner workings of the Brookhaven collider and worked hard to help me understand concepts that remain, for me, quite difficult. The visit offered scheduling challenges that were overcome by Barbara's patience and Karen's persistence. Once at Brookhaven, as I talked to Barbara, it became apparent to me that she was searching for the common ground on which we could communicate effectively. Scientists can have a tendency to talk over the heads of nonspecialists or, alternatively, to oversimplify and talk down to nonspecialists, but Barbara found the "just right" Goldilocks zone that stretched my thinking but did not break it. And she did not focus the conversation on her own work and efforts but instead helped me to see that the work at Brookhaven really depends on thousands of people, all working together toward interrelated goals.

Many thanks are due, too, to the entire staff of Brookhaven's Media and Communications Office. Their outreach program includes their website, their frequent publications in the popular press, and their events for visitors. It is one of the finest scientific outreach programs I have ever encountered.

Antonella D'Antoni, who works for the Italian publisher of *Cold*, organized my trip to Rome and was considerate enough to allow me the time I needed to walk in the footsteps of Pliny the Elder and to visit the Vatican in search of the Congregation for the Doctrine of Faith.

John Parsley, my editor at Little, Brown, consistently encouraged me, from our initial conversations about *Heat*, when he offered valuable guidance on how I might structure the book, to the finalization of the manuscript, when he diplomatically offered editorial suggestions. There is an art to working effectively with writers, and John is an artist. Although I cannot pretend to understand the inner workings of the publishing business, I owe gratitude to many others at Little, Brown, including Elizabeth Garriga, Sarah Murphy, Monica Shah, William Boggess, and Peggy Freudenthal.

My agent, Elizabeth Wales, has offered encouragement and advice since I first contacted her about my last book, *Cold*, and throughout the process of writing *Heat*. Writing is about much more than just writing, and an advocate who understands the publishing business is a necessary part of a successful book. Elizabeth's ongoing efforts to educate me and her other writers have not gone unnoticed.

Several people read and commented on drafts or parts of

Heat, including Frank Baker, Lisanne Aerts (my companion and wife), Ishmael Streever (my son), Dennis Haarsagar, Don Rearden, William G. Henry, and Kathryn Temple. Without their feedback, *Heat* would not be the book that it has become.

I talked to many, many others who influenced my work on *Heat*. Some of the encounters were in passing, such as my brief conversation with a train conductor in New York, an impromptu lecture offered by a man running the steam engine at the Drake Well Museum in Titusville, Pennsylvania, and shared memories from a Las Vegas man who had been present for a number of bomb tests in Nevada. I am always amazed by people who are willing to talk to strangers about strange ideas, and I cannot help being grateful to people willing to talk to a writer whom they have never met about a book on a topic such as heat. I often felt as if I were coming across as something of an eccentric, but almost without exception, people took my questions seriously and steered me toward their own favorite stories of high temperatures. Without this sort of openness, a book like *Heat* would be far narrower and less imaginative.

Last, I have to acknowledge the many writers who came before me and whose words influenced this book. There are millions of words on the topic of heat, written by thousands of writers, some famous and others unknown, some still alive and writing and others long dead but not silent. The urge to write, to share ideas with unknown readers, is a valuable human trait and one that I hope lasts throughout whatever remains of the future of humankind.

NOTES

*With a Few References, Definitions, Clarifications,
and Suggested Readings*

PREFACE: A CANDLE'S FLAME

The Chicago firefighter's description of melting skin comes from Steve Delsohn, *The Fire Inside: Firefighters Talk about Their Lives* (New York: HarperCollins, 1996). In the tradition of Studs Terkel, Delsohn offers a collection of firsthand accounts of what it is to be a firefighter, from making it through a very tough entrance examination to saving lives to being badly burned. The interviews hammer home three striking points: firefighters are hardworking blue-collar men and women doing a dirty and demanding job, firefighters are undeniably heroes, and firefighters are just plain interesting characters.

Antoine Lavoisier is also credited with discovering that combustion consumes oxygen. To some working at the time of Lavoisier, cold was also believed to be a subtle fluid, which they called "frigoric," but others believed that cold was nothing more than the absence of caloric. Lavoisier's ideas were described in "Réflexions sur le phlogistique," *Mémoires de l'Académie des Sciences*, 1783, 505–38; reprinted in *Oeuvres de Lavoisier*, vol. 2, trans. M. P. Crosland (1864), 640.

Michael Faraday's 1860 lecture series *The Chemical History of a Candle* has been reprinted many times and remains widely available. For example, the lectures can be found in vol. 30 of *The*

Harvard Classics, edited by Charles Eliot. Faraday is widely admired as one of history's great experimentalists. Albert Einstein kept a photograph of Faraday, presumably to inspire Einstein's own thinking and dedication to knowledge. Like Einstein, Faraday believed in the value of explaining science to nonspecialists. Faraday expressed the following belief in his candle lectures: "Though our subject be so great, and our intention that of treating it honestly, seriously, and philosophically, yet I mean to pass away from all those who are seniors among us. I claim the privilege of speaking to juveniles as a juvenile myself....Though I stand here with the knowledge of having the words I utter given to the world, yet that shall not deter me from speaking in the same familiar way to those whom I esteem nearest to me on this occasion."

William Herschel discovered infrared light in 1800, more than a half century before Faraday's candle lectures. Herschel projected sunlight through a prism, breaking it into a rainbow of its constituent colors and then measuring the temperature of each color, finding the highest temperatures in the red band. He set his thermometer in the colorless zone just beyond the red band and found it to be hotter still.

Harry Houdini's instructions about fire cages and many other tricks come from his *The Miracle Mongers and Their Methods* (New York: Dutton, 1920). The book is widely available in many editions, including a Project Gutenberg edition online. The lengthy subtitle found with some editions says it all: *A Complete Exposé of the Modus Operandi of Fire Eaters, Heat Resisters, Poison Eaters, Venomous Reptile Defiers, Sword Swallowers, Human Ostriches, Strong Men, Etc.* In describing Houdini as a pyromaniac, I do not mean to imply that he burned down buildings or forests, or that he was sexually excited by fires, but merely that he was obsessed with fires and often felt compelled to do things with fire that no normal person would attempt. Whether this was an unhealthy obsession or an admirable quality I leave to the reader's judgment.

The Veritas Society promotes meditation, as well as what it refers to as "the metaphysical arts," including Psi, Magick, and Body Energy. I am not a member, but I have nothing but the best of wishes for people whose viewpoints differ from my own.

The caloric theory of heat (heat as a subtle fluid) and the mechanical view of heat (heat as an expression of molecular motion) were both discussed before Faraday and Tyndall, but the mechanical theory of heat did not completely replace the caloric theory until after Faraday's death. John Tyndall's *Heat: A Mode of Motion* (New York: D. Appleton), is available in both paper and electronic forms. My source throughout this book is the fourth edition, first published in 1875. The first edition was published in 1863.

CHAPTER 1: RAVING THIRST

Constantin Yaglou and David Minard were the doctors whose work eventually led to the Wet Bulb Globe Temperature as a measure of heat impact on humans. Their report was titled "Prevention of Heat Casualties at Marine Corps Training Centers" (Washington, D.C.: Office of Naval Research, Physiology Branch, 1956). The report comes with a cover sheet that says, in large bold font, at both the top and the bottom of the page, "UNCLASSIFIED." The yellow and red flag condition descriptions come directly from their report. The black flag condition description comes from Marine Corps Order 6200.1E W/CH 1, dated June 6, 2002.

The Wet Bulb Globe Temperature is typically measured as WBGT = $0.7T_w + 0.2T_g + 0.1T_d$, where T_w is the wet-bulb temperature, which captures the cooling effect of evaporation, T_g is the globe thermometer temperature, or the black globe thermometer temperature, which captures heat from exposure to the sun, and T_d is the dry-bulb temperature, or air temperature in the shade. T_w decreases in dry air and in a breeze, and T_g decreases under overcast skies.

Claude Piantadosi offers good technical summaries of dehydration and heatstroke in *The Biology of Human Survival: Life and Death in Extreme Environments* (New York: Oxford University Press, 2003).

The fatty sugars that are released from the intestine as a response to overheating are lipopolysaccharides, or lipoglycans. They are commonly found in Gram-negative bacteria. Once in the blood, they act as toxins and trigger an immune response, which can include a fever.

In their article "Heat Stroke," *New England Journal of Medicine* 346 (2002): 1978–88, Abderrezak Bouchama and James Knochel define heatstroke as "a form of hyperthermia associated with a systemic inflammatory response leading to a syndrome of multi-organ dysfunction, in which encephalopathy predominates."

Casual exploration of abandoned mineshafts in Death Valley—or, for that matter, anywhere—is not recommended. Possible dangers include cave-ins, bad air (concentrations of methane, carbon dioxide, or other gases), falls down vertical shafts, presence of old and unstable explosives, and rattlesnakes. While the borax shafts that I found in Death Valley did not seem dangerous, warning signs and discussions with rangers suggest that it is possible and even easy to find shafts that pose real risks.

Lavoisier summarized his thoughts about respiration in "Experiments on Animal Respiration and the Changes Occurring When Air Passes through the Lungs," read to the Académie des Sciences in 1777.

Antoine Lavoisier's comments about sweat come from vol. 2 of *Oeuvres de Lavoisier*, a collection of Lavoisier's correspondence, originally published in 1864 and later translated by M. P. Crosland. The collection is widely available electronically.

W. J. McGee's "Desert Thirst as Disease," *Interstate Medical Journal* 8, no. 3 (March 1906), describes Pablo Valencia's ordeal. The paper is fascinating not only for its content but also for its style— a narrative style that would not be published in a scientific journal today. For such an amazing piece of work on a topic that should interest anyone visiting deserts, McGee's paper is surprisingly difficult to find outside academic libraries. The paper was reprinted in *Journal of the Southwest* 30, no. 2 (1988), and it may be possible to obtain copies for a small fee from the Southwest Center at the University of Arizona. It can be found online at JSTOR, for those with JSTOR access, and at least occasionally on other sites, such as http://eebweb.arizona.edu/courses/ecol414_514/readings/ thirst.pdf, where it is accompanied by a historical commentary called "W. J. McGee's 'Desert Thirst as Disease,'" written by Bill Broyles, B. W. Simons Jr., and Tom Harlan, which originally appeared in the *Journal of the Southwest* side by side with the reprint of McGee's original article. According to Broyles and his coauthors, McGee's article was first presented to a group of Missouri doctors in 1906, after which it was more or less lost for decades and then circulated as photocopies before finally being reprinted in the *Journal of the Southwest*. It has been especially popular with desert rescue organizations in part because it extends the time of viable survival (and therefore the time that should be allocated to a search).

McGee's "Thirst in the Desert," *Atlantic Monthly* 81 (1898): 483–88, begins with a seven-word sentence: "It is not a pleasant thing, thirst." This article, written seven years before McGee's encounter with Pablo Valencia, is based in part on his own experiences with thirst. The article describes the stages of death from lack of water, and McGee establishes his authority by describing himself as "one who has run the gauntlet two thirds through." He writes of stumbling on a spring and soaking in it for an hour before he regained the ability to swallow. "Then," he wrote, "despite a half-inch cream of flies and wasps, squirming and buzzing above and macerated into slime below, I tasted ambrosia!" His later article, in 1906, used the encounter with Pablo Valencia as a springboard for further discussion of the stages of dehydration.

Prospectors sometimes say that desert trumpets are a sign of gold, but really they are a sign of mineral soil, which might or might not contain gold.

To make mesquite pinole, the mesquite seedpods should be collected in September or October. The pods can be dried in the sun or in an oven until they become crumbly. Traditionally, the next step would involve grinding, probably using stones, one of which would probably be worn into a bowl or at least a depression and might be called a metate, but today's cooks can use blenders or food processors. The pinole can then be used just as other flours are used, but most recipes suggest mixing it with modern white flour, presumably to kill the mesquite taste, which does not suit the modern palate.

The California professor who tethered lizards to watch them die in the heat was Raymond Cowles, famous for his work on thermoregulation. He worked in the western Sonoran Desert near Indio, California. Written with Elna Bakker, his book *Desert Journal: A Naturalist Reflects on Arid California* (Berkeley: University of California Press, 1977) describes some of the work, including the personal revelation about his own ability to survive while dozens of reptiles were killed.

W. Hale White's "A Theory to Explain the Evolution of Warm-Blooded Vertebrates," *Journal of Anatomy and Physiology* 25, pt. 3 (April 1891): 374–85, is available online at www.pubmedcentral. nih.gov/articlerender.fcgi?artid=1328175.

Many references provide information about warm-blooded fish. One is Stephen Katz's "Design of heterothermic muscle in fish," *Journal of Experimental Biology* 205 (2002): 2251–66. Katz examines the role of warm-bloodedness "in fish whose swimming performance is considered elite," including some sharks, tuna, and billfish.

The arrangement of blood vessels in tuna allowing warm blood from muscles to raise the temperature of cold blood from gills is the same as that found in whales, birds, and some other animals, the so-called *rete mirable*, or "miracle network," a countercurrent exchange system much like that which would be designed by an engineer to maximize the use of waste heat in certain kinds of motors and generators.

The statistics on heat-related deaths in America come from the Centers for Disease Control. See www.cdc.gov/climatechange/effects/heat.htm. A growing medical literature warns of the dangers of hot-weather events, documents impacts from past hot-weather events, and offers suggestions to minimize potential dangers. For example, J. Semenza and coauthors, in "Excess Hospital Admissions during the July 1995 Heat Wave in Chicago," *American Journal of Preventative Medicine* 6, no. 4 (1999): 269–77, write: "The majority of excess hospital admissions [during a heat wave] were due to dehydration, heatstroke, and heat exhaustion, among people with underlying medical conditions. Short-term public health interventions to reduce heat-related morbidity should be directed toward these individuals to assure access to air conditioning and adequate fluid intake."

It is possible under extreme conditions to lose as much as 0.66 gallons of water per hour as sweat, but only for a short time. Sustained water loss seldom exceeds 0.20 gallons an hour.

K. N. Moss's paper was "Some Effects of High Air Temperatures and Muscular Exertion upon Colliers," *Proceedings of the Royal Society of London, Series B, Containing Papers of a Biological Character* 95, no. 666 (1923): 181–200. The work described in the paper was undertaken by Moss when he was working as a Tyndall Research Student. The refusal to drink when working in hot conditions—such as those found in deep coal mines—was later called "voluntary dehydration."

The various stages of heat illness are presented in different ways by different "experts," usually without reference to W. J. McGee's classification system. However, all these classification systems are similar. A typical classification system might describe six stages: heat stress (fatigue, dizziness, some swelling of fingers or toes, stumbling, heat rash, headache, thirst), heat fatigue (burning sensation, excessively dry lips, cracking of lips, dry mouth), heat syncope (body temperature rises, pale skin, some difficulty talking, mind starts to wander, rapid heartbeat), heat cramps (muscle cramps, stomach cramps, clumsiness, muscle pain), heat exhaustion (high fever, nausea, loss of will, pounding heart, shallow and rapid breathing, skin cold to touch, possibility of heart attack, loss of consciousness, tunnel vision, hearing anomalies, mild hallucinations), and heatstroke (sweating stops, body temperature spikes, blushing as blood is sent to skin, burst blood vessels in eyes, hypersensitivity to touching or rubbing sometimes leads to abandonment of clothes, severe hallucinations, irrational digging, organ failure). However, symptoms vary from person to person and case to case. It is possible to suffer from heatstroke even when adequate water is available, and to reach the point of heatstroke very quickly—for example, through forced exercise.

People continue to die of heat and thirst in the United States. In 2009 a female prisoner left in an outdoor holding cell in Arizona died. Also in 2009, an eleven-year-old boy died when he and his mother's car got stuck in sand near Death Valley (the mother was severely dehydrated but survived). In 2011 two farm supervisors in California negotiated plea bargains over charges related to the heat death of a young farmworker. Every year, weekend vacationers die in the desert, often within a mile or two of a road and sometimes during relatively short forays—for example, during hikes as short as four hours. Aside from the elderly and infirm, perhaps the single largest group of people likely to die from heat and thirst consists of illegal immigrants crossing the border from Mexico into the United States. Luis Alberto Urrea's *The Devil's Highway* (New York: Little, Brown, 2004) offers a gripping account of twenty-six illegal immigrants who crossed the desert in the same general area that Pedro Valencia had wandered. Fourteen of them died.

In keeping with changing times and changing threats, the Nevada Test Site was renamed the Nevada National Security Site in 2010.

The Death Valley forty-niner who wrote about Owen's Lake was the Reverend John Wells Brier, who crossed the desert with his wife and family. His words are quoted in Charles W. Meier, *Before the Nukes: The Remarkable History of the Area of the Nevada Test Site* (Lansing Publications, 2006). Brier was not popular among the other forty-niners. William Manly's famous memoir comments on Brier. From the Manly memoir, first published in 1886 as *From Vermont to California*, and later, in 1894, supplemented by information provided by other survivors of the Death Valley crossing, as *Death Valley in '49*, and readily available today as an electronic book through both Google Books and Project Gutenberg: "Some were quite sarcastic in their remarks about the invalid preacher [Brier] who never earned his bread by the sweat of his brow, and by their actions showed that they did not care very much whether he ever got through or not." Later, in Death Valley, Manly went on to save the lives of his fellow travelers, finding a way out of the valley, finding water, and then returning for the others, who, close to death, were lying under their wagon, surrounded by dead oxen. According to Manly and others, Death Valley was named when one of the forty-niners, upon leaving the valley, turned back and said, "Goodbye, Death Valley!" In fact, only one of the forty-niners died in Death Valley itself, though the entire party certainly could have died there and seemed to have survived only by a combination of luck, trail sense, sheer toughness, and, for a few of them, Manly's willingness to return with water and aid. Another source of information on the forty-niners' crossing of Death Valley is Louis Nusbaumer's much-lesser-known *Valley of Salt, Memories of Wine: A Journal of Death Valley, 1849* (Friends of the Bancroft Library, University of California, Berkeley, 1967). Nusbaumer was with Manly but spoke little English. He kept his diary in German and wrote across adjacent pages, some of which were subsequently lost, so that Nusbaumer's surviving work includes partial sentences. From a left page, with its matching right page missing: "When I [missing words] I almost died of thirst [missing words]."

Pliny's description of fever comes from *The Natural History of Pliny, Translated, with Copious Notes and Illustrations, by the Late John*

Bostock, M.D., F.R.S., and H. T. Riley, Late Scholar of Clare Hall, Cambridge, vol. 2 (London: George Bell and Sons, 1890). The book is available electronically at http://books.google.com. Bostock, one of the volume's two translators, died with the fever that accompanies cholera in 1846. Other editions of Pliny's work are available from various sources, such as http://penelope.uchicago.edu/Thayer/E/Roman/Texts/Pliny_the_Elder/home.html.

Pliny also recognized that malaria would not begin in winter, but he did not link this recognition to the scarcity of mosquitoes, and he certainly did not understand that mosquitoes were the vector that carried malaria to humans. Pliny was not the first to recognize the periodic fevers associated with malaria—the ancient Chinese recognized this phenomenon by 2700 BC.

Carl Wunderlich is also credited with introducing the routine use of temperature records in hospitals. His *Medical Thermometry and Human Temperature* (New York: William Wood, 1876) was translated into English by E. Seguin, M.D. The medical thermometers of that time required a skillful operator and were far less accurate than those used today.

Robert Fortuine's *Chills and Fever: Health and Disease in the Early History of Alaska* (Anchorage: University of Alaska Press, 1992) offers a fascinating history of the spread of disease through native populations that lacked immunity to European germs and viruses. Fortuine attributes the description of "the man and his wife and three or four children" to A. Parodi, who was writing about the village of Holy Cross, an inland community with a population of about two hundred people in 2000. James W. Van Stone's *Ingalik Contact Ecology: An Ethnohistory of the Lower-Middle Yukon* (Chicago: Chicago Field Museum of Natural History, 1979) also offers the Parodi quote and references a document called "Process of the Plague at Holy Cross Mission, Alaska," written in 1900.

The description of Kuda Box's firewalking came from a chapter called "Science Solves the Fire-Walk Mystery" in Harry Price, *Fifty*

Years of Psychical Research (New York: Longmans, Green, 1939). "The experiments with Kuda Box," Price wrote, "were held under the auspices of the University of London Council for Psychical Investigation and proved very interesting and instructive." Price investigated various other paranormal phenomena. In addition to the chapter on firewalking, his book includes chapters on "mental mediums," "ESP," and "the mechanics of spiritualism." The book remains widely available in print form and can be viewed in part on Google Books. Kuda Box went on to perform firewalking in front of New York's Rockefeller Center in a 1938 demonstration set up by Robert Ripley, creator of *Ripley's Believe It or Not.*

The film clip of Tolly Burkan walking through fire was recorded as part of Michael Shermer's series *The Unexplained*. The clip, available on YouTube, is called "Michael Shermer Firewalking across Hot Coals."

The Firewalking Institute of Research and Education is based in Texas. From their website: "Fire has been around since our beginnings as humans and holds us in thrall with conflicting emotions. Fire is a comfort, a warm home, a tool for empowerment in more ways than one and a source of fear of physical contact with it....Now you can learn another use of fire by walking on it. Sound crazy? I guarantee you it's not. Learn how something as unrecognized as fire can change your perception of life and the world around you through the act of confronting and conquering your fear, building a coherent bond between you and co-workers, team mates, your family, your friends, and even yourself. Believe in the power that you possess to walk on fire and face down any challenge life can throw at you." The firewalking instructor whom I interviewed repeatedly by telephone was Matt Carr, whose insights and enthusiasm were nothing short of inspirational.

The story of the USS *Indianapolis* is well known. Although about one hundred men survived without freshwater for four days, it is believed that an equal number died from dehydration during those four days. Woody Eugene James's story, in his own words, can be found in full at www.ussindianapolis.org/woody.htm.

CHAPTER 2: UNMANAGED FIRE

The photographer whose house burned down is David Gala. His outlook on the Tea Fire, which destroyed most of his possessions, including hundreds of photographs taken all over the world, was admirably philosophical, without a hint of bitterness. Some of his fire artifact photographs can be viewed at www.teafireartproject.com. From the Tea Fire Project website: "On the night of November 13, 2008, fire swept through the hillsides above Santa Barbara, burning in its wake more than 200 homes. In its aftermath, as we searched through the rubble and the ashes for anything to salvage from among the emptiness and desolation, we were struck by the discovery of objects and artifacts unexpectedly transformed by the fire into pieces of art."

The twenty-foot-tall fungus or lichen is *Prototaxites*. It is usually thought of as a fungus, although it may have supported symbiotic algae, which would make it a lichen. The tree-sized fossil remains probably represent fruiting bodies, analogous to mushrooms in today's world.

The quotation regarding Michael Faraday's burning of lycopodium comes from his lecture series *The Chemical History of a Candle* (1860; reprinted in *The Harvard Classics*, vol. 30, ed. Charles Eliot). Faraday, in a footnote appended to his lecture, pointed out that the lycopodium powder was "found in the fruit of the club moss (*Lycopodium clavatum*)." In the same footnote, he mentioned the use of the powder in fireworks.

Joseph Fourier's key papers on the temperature of the earth are "Remarques générales sur les temperatures du globe terrestre et des espaces planétaires," *Annales de Chimie et de Physique* 27 (1824): 136–67, translated by E. Burgess as "General Remarks on the Temperature of the Terrestrial Globe and the Planetary Spaces; by Baron Fourier," *American Journal of Science* 32 (1837): 1–20;

and "Memoir sur les temperatures du globe terrestre et des espaces planétaires," *Mémoires de l'Académie Royale des Sciences de l'Institut de France* 7 (1827): 570–604.

There is, of course, nothing new about chaparral fires. From the *San Bernardino Daily Courier* in 1889: "During the past three or four days destructive fires have been raging in San Bernardino, Orange, and San Diego.... It is a year of disaster, wide spread destruction of life and property—and well, a year of horrors."

The scientist whom I describe as a mapper is Professor Dar Roberts of the University of California, Santa Barbara. Among his many fine qualities, Dr. Roberts is a dedicated teacher who is clearly excited to talk about his work with students—both those enrolled in his classes and those who show up in the guise of curious writers. He is known for his work using satellite imagery to better understand what is happening here on earth, mapping everything from fires to deforestation to roads.

The role of pine beetle impact in fires has been studied in part on the basis of satellite imagery. www.nasa.gov/topics/earth/features/beetles-fire.html summarizes one study.

CalFire News posts succinct and interesting fire notes on a website in real time to keep the public informed about fires. Not surprisingly, most wildland fires and fire responses can be characterized as confusing. *CalFire News* carries the following note in the banner: "INFORMATION PROVIDED RAW MAY NOT BE TIMELY OR UPDATED REGULARLY." *CalFire News* is not affiliated with CalFire, the largest fire department in California.

There are many accounts of "mopping up" in the literature of fires and firefighting. One, by Stephen J. Pyne, is especially interesting, in part because Pyne is best known for his academic accounts of fire history. It is a pleasure to come across a writer known for his academic work only to find that he led a previous life as a firefighter and describes this life in the first person. There is no glory in mopping up. It is dirty, hot, menial labor. Most would agree that Pyne made the right choice in switching to academics.

In "Historical Santa Barbara Up in Flames: A Study of Fire History and Historical Urban Growth Modeling," a paper presented to "The Fourth International Conference on Integrating GIS and Environmental Modeling (GIS/EM4): Problems, Prospects and Research Needs," Banff, Canada, September 2000, Noah Goldstein, Jeannette Candau, and Max Moritz used map data to estimate the size of well-known fires. The total area burned in historical fires that touched the greater Santa Barbara urban area from 1955 through 1979 (the Refugio Fire, the Polo Fire, the Coyote Fire, the Romero Canyon Fire, the Sycamore Canyon Fire, and the Eagle Fire) was 69,697 hectares, or about 172,000 acres. The paper can be found online at www.geog.ucsb.edu/~kclarke/ucime/banff2000/530-ng-paper.htm.

For an example of a simple fire model depicting wildland fires over a long period, see http://schuelaw.whitman.edu/JavaApplets/ForestFireApplet. Cells in the model change color as they burn and again as they burn out. In the model, the ignition source of each fire is lightning. When the model starts, trees grow quickly, turning cells green. As trees mature, lightning strikes start fires. Burning cells are white. As a fire spreads, the burned-out cells behind it turn brown. Trees regrow in the burned-out cells, and they eventually reburn. Obviously models as simple as this one are not realistic as predictive tools but are intended to illustrate trends. Perhaps less obviously, more complex models are also unrealistic. Fires will burn as fires burn, and to date no one has come up with a model—or the data to populate the model in a manner that captures real-world variability—that reliably predicts actual fire behavior.

The quote by G. W. Craddock comes from his master's thesis, "The Successional Influences of Fire on the Chaparral Type" (University of California, Berkeley, 1929). The thesis is quoted in Mark I. Borchert and Dennis C. Odion, "Fire Intensity and Vegetation

Recovery in Chaparral: A Review," in *Brushfires in California Wild-lands: Ecology and Resource Management,* ed. J. E. Keeley and T. Scott (Fairfield, Wash.: International Association of Wildland Fire, 1995). The paper also includes plots showing ground temperature at different points in a chaparral fire.

The firefighter who showed me the Spanish Ranch fire and other fire sites was Louis De La Rosa. Louis has also talked to other writers, including Joseph N. Valencia, author of *Area Ignition* (see next note), who quoted Louis's remarks about his own training ride, or staff ride, at Spanish Ranch: "When I first came on this staff ride, I didn't know the entire story. This fire happened in my backyard. I take it very seriously."

The information about the Spanish Ranch Fire came from two sources: a *Staff Ride Students Guide,* printed by the CDF San Luis Ranger Unit as a training document, and Joseph Valencia's book *Area Ignition: The True Story of the Spanish Ranch Fire* (Blooming-ton, Ind.: AuthorHouse Press, 2009). The phrase "area ignition" is sometimes used to refer to areas where heated gases released by trees and brush suddenly ignite, making it appear as though the sky is on fire. The photograph of a charred Nomex shirt appears in the *Staff Ride Students Guide.* Ed Marty's photograph, showing Marty in a Smokey Bear suit, appears in Valencia's book.

Nomex was developed by DuPont in the 1950s. It is in some ways similar to nylon (which was also developed by DuPont, in the 1930s) but more rigid, durable, and extremely fire resistant. It is used by firefighters, race car drivers, military pilots, and oil field workers.

The discussion of Gifford Pinchot is for the most part based on Timothy Egan's wonderful *The Big Burn: Teddy Roosevelt and the Fire That Saved America* (New York: Houghton Mifflin Harcourt, 2009). Egan's book presents the early history of the National Forest Service and the fire itself, in part through descriptions of Pinchot's role, but also through the roles of various firefighters. The firefighters included recent immigrants to the United States who had no experience in the wilderness.

The full title of Robert Burton's 1621 *The Anatomy of Melancholy* is *The Anatomy of Melancholy, What it is: With all the Kinds, Causes, Symptomes, Prognostickes, and Several Cures of it. In Three Maine Partitions with their several Sections, Members, and Subsections, Philosophically, Medicinally, Historically, Opened and Cut Up*. Although the book is renowned as a unique work of literature and is sprinkled with passages of genius, most readers today would probably find the book even more tedious than its full title.

Edward Abbey's remarks about Smokey Bear come from his book *The Journey Home* (New York: Penguin, 1991). In one passage, Abbey speaks out against the metric system. "The Park Service," he writes, "no doubt at the instigation of the Commerce Department, is trying to jam the metric system down our throats, whether we want it or not. We can be sure this is merely the foot in the door, the bare beginning of a concerted effort by Big Business—Big Government (the two being largely the same these days) to force the metric system upon the American people. Why? Obviously for the convenience of world trade, technicians and technology, to impose on the entire planet a common system of order. All men must march to the beat of the same drum, like it or not." The metric system has fallen by the wayside in the United States, but even without it the big business–big government partnership seems to have thrived since Abbey wrote *The Journey Home*.

There are many estimates of the degree to which fires are started by lightning as opposed to humans playing with matches or otherwise providing a source of ignition. Abbey stated that 90 percent of western forest fires were ignited by lightning, but I could not find his source for this statistic. Sources on this matter will vary from place to place and year to year, and while Abbey's estimate may have been high, lightning strikes are important sources of ignition for wildfires in many areas, especially where dry lightning—lightning from storms that do not shower the ground with rain—is common. Most reports of fire ignition sources in the American West suggest that well over half of all fires are ignited by lightning.

For official information about Smokey Bear, see the government publication *Smokey Bear Guidelines*, dated March 2009. The guidelines explain how Smokey can and cannot be used. From page 3 of the guidelines: "Today, Smokey Bear is a highly recognized advertising symbol and is protected by Federal law. (PL 82-359, as amended by PL 92-318)." The guidelines are available online at www.smokeybear.com/downloads/Smokey_Bear_Guidelines.pdf.

There are many sources of information about fire injuries and deaths. The U.S. Fire Administration, which is part of the Department of Homeland Security's Federal Emergency Management Agency (FEMA), publishes profiles of fires in the United States. At least thirteen editions are in circulation. The thirteenth edition was called *A Profile of Fire in the United States, 1992–2001* (published by FEMA and available online). Among other things, it includes statistics for firefighter deaths. In 2001, 449 firefighters died, with 344 dying in the World Trade Center and 102 in other situations. "As in all previous years," the report says of the 102 who died in fires other than the World Trade Center, "the most frequent cause of deaths was stress or over exertion." Heart attacks and strokes were the most common cause of death. Fourteen firefighters died during training exercises.

Wall's theory, as it is called in some sources, is more appropriately called gate control theory. Many people have worked on gate control theory, but Patrick Wall and Ronald Melzack are generally credited as its originators. Their seminal article "Pain Mechanisms: A New Theory" was published in *Science* 150, no. 3699 (1965): 971–79. The authors later acknowledged that many of their premises were incorrect, yet the general ideas behind gate control theory continue to influence pain research and treatment. Comments by Wall and Melzack shed an interesting light on the inner workings of science. From Melzack's oral history: "So in the course of our talking I said to Pat, 'You know, you and I think a lot alike about a lot of things. Why don't we write a paper together?' So we wrote a paper that was published in *Brain* in 1962. And we struggled with that paper, putting it all together, and it was certainly jointly done all the way through. I think three people read the paper. So we began to write [a second] paper and sending drafts back and forth—I'd bring them down, we would argue, and so on, and then at some stage, we began to organize the paper into components, and the main, the gate control theory got invented. Anyway, I suggested that we really aim for the top and try *Science* and see what the hell happens—the worst that will happen is to get rejected. It got accepted. We were astounded. Well, so, then you know the rest because some people loved it and most people hated it." From Wall's oral history: "At this time, with a completely different background, there was Ron Melzack, with whom I've really never worked in my life, we'd only got to talking. You ask about the gate control theory, which is 1965 as published, if you read what we'd published certainly three years before, it says exactly the same thing in it. And we tossed a coin, and published essentially exactly the same paper, only as Wall and Melzack (1962) rather than Melzack and Wall (1965), and it was utterly ignored. And then we put out the *Science* paper. And as you see, if you read this, we simply tried to bring together everything that we knew and what was in the literature at the time, knowing very well that we could be wrong, and certainly in the details." Both oral histories are deposited with the UCLA biomedical library.

The realization that the brain is involved in the perception of pain seems intuitive today, but that has not always been the case. Not surprisingly, pain has been attributed to evil spirits or as punishment for some moral or spiritual shortcoming. Perhaps more surprisingly, the Greeks, under the sway of Hippocratic medicine, saw pain as an expression of an imbalance between the four humors of yellow bile, black bile, phlegm, and blood, each of which had its own qualities and an association with seasons and the elements of fire, earth, water, and air, respectively. It was not until the seventeenth century that work by René Descartes attributed the sensation of pain to the brain. A famous sketch from Descartes's work shows a naked boy with his foot next to a fire and a link between the brain and the foot running inside the boy's body. Descartes saw this as a one-way flow of information. He described nerve impulses as "fast moving particles of fire" and wrote, "The disturbance passes along the nerve filament until it reaches the brain." Today the sensation of pain is seen as something far more complex than this.

The firefighter's remarks about morphine to relieve pain come from Steve Delsohn's aforementioned *The Fire Inside* (New York: HarperCollins, 1996), a compilation of first-person accounts of the lives of firefighters.

Silas Weir Mitchell's *Injuries of Nerves and Their Consequences* (Philadelphia: J. B. Lippincott, 1872) was based on his experience as a Civil War surgeon. During that period, causalgia—pain triggered by the slightest stimulus, usually in a hand or foot that was not directly affected by an actual injury—was sometimes treated by amputation. Causalgia is now known as a type of "complex regional pain syndrome." In his writings, Mitchell also explored the phenomenon of phantom limbs in amputees. Mitchell was both a technical writer and a literary writer, credited with poetry, historical novels, and at least one short story published in the *Atlantic Monthly*.

The article offering case studies of burn victims was I. H. Erb, Ethel M. Morgan, and A. W. Farmer, "The Pathology of Burns: The Pathologic Picture as Revealed at Autopsy in a Series of 61 Fatal Cases Treated at the Hospital for Sick Children, Toronto, Canada," *Annals of Surgery* 117, no. 2 (1943): 234–55.

For more on fire beetles, see Martin Müller, Maciej Olek, Michael Giersig, and Helmut Schmitz, "Micromechanical Properties of Consecutive Layers in Specialized Insect Cuticle: The Gula of *Pachnoda marginata* (Coleoptera, Scarabaeidae) and the Infrared Sensilla of *Melanophila acuminata* (Coleoptera, Buprestidae)," *Journal of Experimental Biology* 211 (2008): 2576–83. For a more accessible summary of this article, see Kathryn Phillips, "Beetles 'Hear' Heat through Pressure Vessels," *Journal of Experimental Biology*, August 8, 2008, i–ii. Louise Dalziel's short article "Jewel Beetle Flies into the Inferno," *BBC*, March 20, 2005, also offers a good summary, available online at http://news.bbc.co.uk/go/pr/fr/-/2/hi/science/nature/4362589.stm.

The government publication on the evolutionary insignificance of fire for wildlife came from Ronald D. Quinn, "Habitat Preference and Distribution of Mammals in California Chaparral, Research Paper PSW-RP-202" (U.S. Department of Agriculture, Forest Service, Pacific Southwest Research Station, Berkeley, Calif., 1990).

Charles E. Little's article "Smokey's Revenge," *American Forests* 99 (May–June 1993): 24–25, 58–60, offers a short but thoughtful discussion of Smokey's role in the history of American wildfires, both pro and con. Another interesting short article is Jim Carrier's "An Agency Icon at 50," *High Country News*, October 3, 1994.

The Stephen J. Pyne quote about the Yellowstone fire came from his article "The Summer We Let Wild Fire Loose," *Natural History*, August 1989, 45–49.

The photographer's mother is Betsy Gallery. The fire mural that she created is on display at the Santa Barbara County fire headquarters, a very suitable home for such a work. While much of her work requires large spaces (such as public buildings) for display, she also creates smaller murals suitable for display in typical homes. Her website, www.elizabethgallery.com, correctly and succinctly says that her "studio resembles an archaeological dig as much as an artist's workshop." The website includes a photograph of the completed and quite amazing eight-by-four-foot Santa Barbara fires mural.

CHAPTER 3: COOKED

The Rio Earth Summit is known by many names. Officially it was "The United Nations Conference on Environment and Development (UNCED), Rio de Janeiro." The full text of the convention can be found in seven languages at http://unfccc.int/essential_background/convention/background/items/2853.php. Most of the nineteen thousand environmentalists who attended were there primarily to participate in the Global Forum, sometimes referred to as "the world's fair of environmentalism," which was set up in a downtown park in Rio de Janeiro well removed from the conference delegates and their official discussions.

John Tyndall's paper was "On the Absorption and Radiation of Heat by Gases and Vapours, and on the Physical Connexion of Radiation, Absorption, and Conduction: The Bakerian Lecture," *London, Edinburgh, and Dublin Philosophical Magazine and Journal of Science*, September 1861. The paper offers tremendous insight into the amount of research required to probe the workings of nature. It also includes a wonderful diagram of the apparatus that Tyndall used in these experiments. The paper can be found online at http://onramp.nsdl.org/eserv/onramp:1657/n3.Tyndall_1861corrected.pdf. Tyndall's heat source, the cubical bucket, was a Leslie cube, developed by John Leslie, who used the cube to experiment with radiant heat in 1804. The cube has one side of highly polished metal, two sides of copper, and one side painted black. When filled with boiled water, most of the radiant heat is transmitted through the black side of the cube.

Aside from the information attributed to Tyndall's 1861 paper, all the other quotes and thoughts attributed to Tyndall come from his aforementioned book *Heat: A Mode of Motion* (New York: D. Appleton, 1875).

John Slusher's article "Wood Fuel for Heating" (University of Missouri Extension Service, March 1985) provides a table listing the British thermal unit (Btu) yield of various kinds of firewood per cord. A cord of red cedar burns to produce 18.9 million Btu. One gallon of gasoline yields about 115,000 Btu of heat when burned. Therefore burning one cord of red cedar produces the same amount of heat that would be produced by burning about 164 gallons of gasoline.

John McPhee's essay "Firewood," first published in the *New Yorker* in 1974 but reprinted in *Pieces of the Frame* (New York: Farrar, Straus and Giroux, 1975), provides an amazing account of the use of firewood in New York during the energy crisis of the 1970s.

Benjamin Franklin described his stove in "An Account of the New Invented Pennsylvanian Fire-Places," a few-thousand-word essay with a seventy-word subtitle. Just below the subtitle, he wrote, "Printed and Sold by B. Franklin. 1744." The essay is readily available online. The main competitor for the Franklin stove was the Rumford fireplace—essentially a modified standard brick fireplace, but innovative in its time because Rumford's seemingly minor addition of a few bricks along the walls and at the base of the chimney improved the updraft, guiding smoke up the chimney and in so doing bringing more oxygen across the fire. Count Rumford, also known as Benjamin Thompson, was the same man whom Tyndall described boiling water with a cannon borer and who later married Lavoisier's widow. With Sir Joseph Banks, Rumford founded the Royal Institution, which later employed Michael Faraday. And, importantly, he is credited with the invention of the coffee percolator.

Franklin wrote of turning down the opportunity to patent his stove in his autobiography, which he wrote, at least ostensibly, for his family. *The Autobiography of Benjamin Franklin* remains available. My copy is an electronic version originally published in 1909 by P. F. Collier and Son, New York. One can only imagine that Franklin, with thoughts of lightning and his famous kite, would be delighted to know that his book can now be read electronically. He would probably comment on the benefits of electronic publishing, not the least of which might be the savings in trees, which could better be used in his stoves. And, of course, he would scrap his presses and replace them with computers.

The burned chemistry textbook was Francis A. Carey's *Organic Chemistry* (New York: McGraw-Hill, 1987). Carey did an excellent job with this textbook, but in the end it was still a textbook, and for that matter a very old textbook. Before burning the book, I found stuffed within its pages twenty sheets of unlined paper on which a younger version of myself had scrawled notes in the code of organic chemistry—formulas with arrows and triangles and lines representing bonds. I recognized my handwriting but understood none of it, a reality for which neither Carey nor the professors who tried to teach me organic chemistry can be blamed.

Henry David Thoreau's remark regarding the ability of wood to warm twice comes from his book *Walden*, first published in 1854. In *Walden*, Thoreau wrote often of fire. Among other things, he wrote: "Food may be regarded as the Fuel which keeps up the fire within us—and Fuel serves only to prepare that Food or to increase the warmth of our bodies by addition from without."

The details of Zoroaster's life—including exactly when he lived—are lost.

Brenda Fowler's *Iceman: Uncovering the Life and Times of a Prehistoric Man Found in an Alpine Glacier* (University of Chicago Press, 2001) offers a detailed account of the Iceman's discovery and exhumation. The body of the Iceman was mummified under the ice—that is, both frozen and dried out. He was probably in his mid-forties when he died. In life, he probably stood about five feet, five inches tall and probably weighed about 110 pounds. Gut contents suggest that he may have eaten bread just before he died, which means he had access to some form of baking. He carried two birch bark pouches, one of which included a flint and pyrite for striking sparks. The museum housing the Iceman is in Bolzano, Italy. The Iceman's frozen carcass can be observed through a small window, and his tools and other belongings are displayed.

The recipes for Asháninka dishes come from the website "Cooking the Native Peoples of Brazil" (http://hubpages.com/hub/Cooking-the-native-peoples-of-Brazil), which itself references an apparently inaccessible book called *Encyclopedia of the Forest: The High Jurua; Practice and Knowledge of the People*.

Frances Burton's *Fire: The Spark That Ignited Human Evolution* (Albuquerque: University of New Mexico Press, 2009) describes interactions between various primates and humans. "The acquisition of fire," she writes, "was the engine that propelled the incredibly fast evolution of humans." She sees the control of fire not as a by-product of human evolution but as "a, or maybe the, major contributor to the manifold changes that made us human."

Richard Wrangham's *Catching Fire: How Cooking Made Us Human* (New York: Basic Books, 2009) outlines the theory that cooking led to rapid human evolution. While Frances Burton believes that control of fire—which included cooking as one of its many aspects—accelerated human evolution, Wrangham believes that cooking was the primary advantage offered by the control of fire. To Wrangham himself, the theory is so obvious, in retrospect, that he seems a bit embarrassed to present it as a novel idea. "What is extraordinary about this simple claim," he writes, "is that it is new." Burton cites several of Wrangham's articles in her book *Fire: The Spark That Ignited Human Evolution* (see previous note) but seems to see cooking as only part of the formula, focusing more broadly on the many advantages offered by the control of fire.

Jean Anthelme Brillat-Savarin's *The Physiology of Taste*, first published in 1825, remains available from various sources, including as a Penguin Classic. He is credited with the saying "Tell me what you eat, and I will tell you what you are." The foreword in the Penguin edition calls his work "a brilliant treatise on the pleasures of eating and the culmination of his long and loving association with food."

The journalist who interviewed Hérve This was Patricia Gadsby, and her beautifully written article "Cooking for Eggheads" is based on that interview. The article includes her description of This spotting a 64-degree egg sold as a 65-degree egg, but in the text I have converted the temperatures to the Fahrenheit scale. The article was first published in *Discover* and later republished in *The Best American Science Writing*, ed. Gina Kolata (New York: Harper

Perennial, 2007). Hérve This's critique of microwaved beef comes from his book *Kitchen Mysteries: Revealing the Science of Cooking* (New York: Columbia University Press, 2007). Another interesting book by This is *Molecular Gastronomy: Exploring the Science of Flavor* (New York: Columbia University Press, 2006).

The word "cretin," applied to people of low intelligence, originally meant that they were "still human" or "still Christian" and should be treated as such.

Thomas Briscoe's story is based on a statement by Circuit Judge Bazelon and a statement by Circuit Judge Bastian in relation to Petition for Leave to Appeal in Forma Pauperis, 248 F.2d 640, *Thomas E. Briscoe, Petitioner, v. United States of America, Respondent*, misc. no. 855, United States Court of Appeals District of Columbia Circuit, September 20, 1957. Briscoe's case is not resolved in the court records that I obtained.

About 80 percent of children outgrow bed-wetting by age five, but bed-wetting persists in a small percentage of children into their teenage years. I do not know of any statistics supporting Freud's belief in a relationship between bed-wetting and pyromania.

Rosa Briscoe's case is described in 742 F.2d 842, 16 Fed. R. Evid. Serv. 424, *United States of America, Plaintiff-Appellee, v. Rosa Briscoe, Defendant-Appellant*, no. 84-4010, United States Court of Appeals, Fifth Circuit, September 12, 1984. The store that Rosa had burned sold boots. The various defendants involved with the case took boots off the shelves on their way out of the gas-filled store.

Anyone who feels an urge to put a match in a microwave oven should suppress that urge. It is, in fact, potentially dangerous. While my own oven survived, others have not. YouTube and other sources illustrate what happens without the need to ruin your own oven.

The story of Percy Spencer's invention of the microwave oven is available through many sources. Among them is Don Murray, "Percy Spencer and His Itch to Know," *Reader's Digest*, August 1958. "Percy Spencer," the article begins, "was the nosiest man I have ever known." The author then relates a story of Spencer's interest in his (the author's) shoes. Spencer convinced the author to remove one shoe so that he could better examine the stitching. I have not come across any accounts of Spencer experimenting with lit matches exposed to microwave radiation, but it seems like the sort of thing Spencer would have relished. Although he had little formal education, before he retired he was described as "highly technical."

The urge to repeat the experiment of cooking a whole egg (in its shell) in a microwave oven should be suppressed, just as the urge to microwave a lit match should be suppressed. It is possible for the egg to damage the oven. In addition, it ruins the egg. While this may be just the sort of thing that Spencer himself enjoyed doing, the manufacturers of microwave ovens discourage it. As an aside, I also tried a lemon and an apple in the microwave. In both cases, they burst without exploding—a thin slit in the skin opened to release juice and steam. An apple, cooked this way, is too hot to eat in the middle but still cool near the skin.

Frederick Boyle's book was *Adventures among the Dyaks of Borneo* (London: Hurst and Blackett, 1865).

Obviously the manufacturers of disposable lighters would not condone the practice of tossing their product into a campfire. Clearly, burning a disposable lighter is a dangerous and stupid activity.

CHAPTER 4: MY CHILDREN EAT COAL

Peat mining was so important to the area around the Groote Peel that many of the villages end in the syllable *veen*, meaning "peat." There are, for example, Griendtsveen (where Kortooms is buried) and Helenaveen. The villages were probably founded by peat miners.

Toon Kortooms is among Holland's most successful writers. His novel *Beekman and Beekman*, a sort of latter-day Dutch *Tom Sawyer and Huckleberry Finn*, is said to have sold two million copies.

The Drents Museum in Assen, Holland, houses several mummified bog people, including Yde Girl, the Weerdinge Men, and the Emmer-Erfscheidenveen Man. Each is interesting in its own way. For example, the Weerdinge Men—two men, side by side and arm in arm, headless—were first believed to be a man and woman, perhaps man and wife. The museum also houses what is believed to be the world's oldest canoe, the canoe of Pesse, thought to be ten thousand years old. It is so well preserved that charring remains visible, presumably from the original construction, when a combination of fire and stone tools would have been used to carve out the canoe. Yde is pronounced "Ida."

Iron tools—the defining characteristic of the Iron Age—came to different parts of the world at different times. They came first to the Middle East, around three thousand years ago, but spread quickly to India and Europe. Ignoring a few sites in North America where trade routes may have introduced iron tools relatively early, the Iron Age did not spread to the Americas until the time of European colonization. Similarly, the Iron Age came to Australia with European explorers and colonies.

A paper by Keith Branigan, Kevin Edwards, and Colin Merrony, "Bronze Age Fuel: The Oldest Direct Evidence for Deep Peat Cutting and Stack Construction?" *Antiquity* 76, no. 293 (2002): 849–55, describes a stack of peat cut for fuel around 2000 BC in the Outer Hebrides, Scotland. The individual bricks of peat held imprints of fingers and thumbs. The degree to which peat was used as fuel in the Bronze Age and Iron Age is not well known, but it seems obvious that people would have used peat before they would have cut and carried firewood from beyond the immediate vicinity of their homes.

According to Robert Galloway in *A History of Coal Mining* (London: Macmillan, 1882), indoor open hearths were still in use in the late 1800s in Ireland and the highlands of Scotland: "The smoke from fires so situated pervaded the whole apartment, and made its escape by a hole in the roof, or by the doorway—a primitive arrangement to be met with even at the present day in some parts of the Highlands of Scotland and west of Ireland, where peat fuel is still used."

Benjamin Franklin's worries in 1744 about fuel shortages appeared in his self-published essay "An Account of the New Invented Pennsylvanian Fire-Places," and again four decades later in his comments about firewood demand outstripping supply in France, which appeared in his letter to John Ingenhousz, in Vienna, titled "On the Causes and Cure of Smoky Chimneys," written on August 28, 1785. The letter is preserved in *The Works of Benjamin Franklin, Volume VI*, published in Boston in 1884 by Charles Tappan and readily available online.

John Evelyn's *Fumifugium* was reprinted many times—in 1772, 1825, 1930, 1933, 1944, 1961, and 1976. The 1944 printing was by the National Society for Clean Air. I used the 1976 reprint, by the *Rota* at the University of Exeter, which is a facsimile of the 1661 version. It is available online at www.archive.org/stream/ fumifugium00eveluoft#page/n5/mode/2up. The preface offers a short biography of Evelyn, describing him as a "connoisseur of cities." He did not look favorably on London.

Lord Macaulay was a British intellectual and politician active in the first half of the nineteenth century. His comments about chimney taxes come from *The Complete Works of Lord Macaulay, History of England*, which was completed by his sister after his death in 1859. This book has been digitized and is available from Google Books. The Pepysian Library refers to the collections of Samuel Pepys, which, at the time of his death, included thousands of volumes, including copies of his diaries. Per his instructions, the library was bequeathed to Magdalene

College, Cambridge. Pepys was a chronic diarist from 1660 until 1669. His work became an important source of primary information and firsthand impressions, including impressions of the plague in 1665 and the Great Fire of London in 1666, which he described as "a lamentable fire."

The 1842 Parliamentary Commission report—which, in addition to the twelve-year-old girl's testimony, included comments about miners as "dregs" who practiced polygamy—was the "First Report of the Commission for Inquiring into Employment of Children in Mines and Manufacturers, 1842." The report is also quoted by Barbara Freeze in her book *Coal: A Human History*, and she attributed it to J. U. Nef's *The Rise of the British Coal Industry* (London: Frank Cass, 1966).

The account of Charles Dickens's plans to write an article on child labor comes from Peter Kirby, "Early Victorian Social Investigation and the Mines Commission of 1842," Manchester Papers in Economic and Social History, no. 66 (University of Manchester, 2009). Kirby's article references a letter written by Dickens in 1841: "I have made solemn pledges to write about [mining] children."

Charles Dickens's interview with a coal miner was published as "A Coal Miner's Evidence," *Household Words*, December 7, 1850.

When Dickens published *Bleak House*, he was criticized for giving some credence to the existence of spontaneous human combustion. Dickens defended himself in the preface of an 1868 edition of *Bleak House*, claiming, "I do not willfully or negligently mislead my readers. Before I wrote that description, I took pains to investigate the subject." Since then, others have defended the existence of spontaneous human combustion, but if it in fact exists, it occurs very rarely and without any currently plausible explanation. For readers prone to anxiety, there are far better things to worry about than the possibility of spontaneously combusting while reading *Heat*.

John Holland, in *The History and Description of Fossil Fuels, the Collieries, and Coal Trade of Great Britain* (London: Whittaker and Company, 1835), described the draining of the River Garnock. His description was based on "the Scotch newspapers." No one was killed in this incident.

The title of Edward Somerset's *The Century of Inventions* (1655) refers not to one hundred years of inventions but to one hundred inventions. He described both the basic workings of the steam engine and also various tasks to which it could be applied, including the pumping of water. Charles Partington of the London Institution, a science society prevalent in the 1800s, wrote a letter to Doctor George Birkbeck, president of the London Mechanics' Institution, that was included as an introductory note in an 1825 reprint of *The Century of Inventions*: "As a connecting link in the History of the Steam Engine, I know that your attention has been directed to the Marquis of Worcester's *Century of Inventions*, and that its merits were duly appreciated by you at a very early period of Life." Beginning on page 101, this edition includes a description of the history and mechanics of steam engines, along with excellent sketches, including improvements brought about by James Watt. The description argues that Somerset's book contains all the information one would need to understand the workings of Savery's steam engine. This may be true, but it is equally true that Somerset's work lacked the detail that would be needed to build or patent a steam engine.

The full title of the book on steam engines by Dionysius Lardner and James Renwick is *The Steam Engine Familiarly Explained and Illustrated, with an Historical Sketch of Its Invention and Progressive Improvement, Its Applications to Navigation and Railways; with Plain Maxims for Railway Speculators* (New York: A. S. Barnes, 1865). Today this book would interest steam enthusiasts, but it does not convey a simple understanding of steam engines and their workings. Steam engines are not "familiarly explained" in this book, but for highly motivated and patient readers, they are explained well enough through text and illustrations. As an example:

"The fuel is maintained in a state of combustion, on the bars, in that part of the tube represented at D; and the flame is carried by the draft of the chimney round the curved flue, and issues at E into the chimney. The flame is thus conducted through the water, so as to expose the latter to as much heat as possible."

Lardner and Renwick also described concerns regarding traction. An engine might be able to turn wheels, but would the wheels propel the cart or simply spin, to no good effect? They described an approach to traction that modeled itself after the legs and feet of an animal, rather than wheels: "It will be apparent from this description, that the piece of mechanism here exhibited is a contrivance derived from the motion of the legs of an animal, and resembling in all respects the fore legs of a horse. It is however to be regarded rather as a specimen of great ingenuity than as a contrivance of practical utility." It was important for inventors to ascertain "that the adhesion or friction of the wheels with the rails on which they moved was amply sufficient to propel the engine, even when dragging after it a load of great weight."

Stephenson's first train, the *Blücher*, was named after Gebhard Leberecht von Blücher, a Prussian general who helped to defeat Napoleon. Blücher was known for his utter hatred of the French.

Ralph Waldo Emerson's comments about coal as a portable climate come from his 1860 essay "Wealth," which was part of *The Conduct of Life*, a book that today might be sold from the self-help shelves of major bookstores. In this essay, he suggested that we call coal "black diamonds."

Chauncy Harris's description of Europe just after the war comes from "The Ruhr Coal-Mining District," *Geographical Review* 36, no. 2 (1946): 194–221. Harris was a geography professor in Chicago, well known in the field for his work on Soviet geography, an especially important topic during the Cold War. He died in 2003.

CHAPTER 5: ROCK OIL

One important source for information about the history of the oil industry is Daniel Yergin's amazing book *The Prize: The Epic Quest for Oil, Money, and Power* (New York: Simon and Schuster, 1991). Two decades later, *The Prize* is now dated, but it remains amazing in its scope and insightfulness about an extremely complex industry. One can only hope that Yergin will eventually write a sequel to explain how things have changed since *The Prize* was published. Another important source is William Brice's *Myth, Legend, Reality: Edwin Laurentine Drake and the Early Oil Industry* (Oil City, Pa.: Oil Region Alliance of Business, Industry, and Tourism, 2009), which includes many pages of original correspondence and other documents from the time of Drake.

Bissell is often credited with the idea of drilling for oil, but there remains some controversy over this point. Drake may originally have taken the idea to Bissell. Salt wells were common in the area, and it is easy to imagine that drilling for oil would have occurred to more than one person. Similarly, the exact date on which oil was found in the Drake well is a matter of some uncertainty. Some of the early accounts listed dates that did not exist, such as "Saturday August 18, 1859," and "Saturday August 28, 1859," but the eighteenth was in fact a Thursday that year, and the twenty-eighth was a Sunday. Adding to the confusion, neither Drake nor Smith realized they had struck oil until the day after the well had struck oil. In any case, most historical accounts give the date as August 27, 1859, and I have followed that convention here.

Cable drilling or percussion drilling is still used today, but only for shallow wells and more commonly for water wells than for oil or gas wells. Rotary drilling, commonly used in modern oil and gas fields, uses a spinning bit similar to that of drill bits used by carpenters.

Uncle Billy Smith's account of the Drake Well fire comes from an article in the *Titusville Weekly Herald*, January 15, 1880, reprinted in part in William Brice's aforementioned *Myth, Legend, Reality: Edwin Laurentine Drake and the Early Oil Industry*.

The Derrick's Hand-Book of Petroleum (Oil City, Pa.: Derrick Publishing Company, 1898) presents the statistics of the nascent petroleum industry, from 1859 to 1898, with information on exports, field operations, market quotations, and other details. It remains available in both paper and electronic versions.

The well names come from William Wright's *The Oil Regions of Pennsylvania* (New York: Harper and Brothers, 1865), reprinted as an Elibron Classic in 2005. To my knowledge, individual wells are seldom named today, but the practice of naming oil fields continues—for example, Northstar, Liberty, Burger, Klondike, and Alpine are Alaskan oil fields, and Thunderhorse is an oil field in the Gulf of Mexico.

The words in the flier for Kier's Petroleum were preserved in J. T. Henry, *The Early and Later History of Petroleum, with Authentic Facts in Regard to Its Development in Western Pennsylvania* (Philadelphia: Jas. B. Rodgers, 1873). The newspaper article praising rock oil in 1892 was quoted in P. H. Giddens, *Pennsylvania Petroleum, 1750–1872, a Documentary History: Drake Well Memorial Park* (Pennsylvania Historical and Museum Commission, 1947). Mark Twain's words came from a handwritten letter addressed to J. H. Todd, dated November 20, 1905. An electronic copy of the letter is widely available and frequently quoted.

The stories of the *Elizabeth Watts*, the *Zoroaster*, the *Moses* (a third vessel not mentioned in the text), and the *Glückauf* (a fourth vessel not mentioned in the text) are well known and often repeated. Which of these, if any, should be thought of as the first true oil tanker depends on how oil tankers are defined. Unlike the *Elizabeth Watts*, the *Zoroaster* had built-in tankage, but the tanks were not part of the hull of the ship. The *Moses*, another Nobel ship, is sometimes said to be the first tanker to use the inside of its hull as a tank. The *Glückauf* (German for "lucky" or "get lucky") used its hull as a tank but also included extensive plumbing and some safety features common on modern oil tankers. The *Glückauf*, despite its name, ended its life in 1893 on a beach at Fire Island, New York, where sport divers still explore its scattered remains.

The navy's reports on oil were both reprinted in *Pennsylvania Petroleum, 1750–1872*, a collection of various articles and papers compiled by Paul Giddens and published in 1947 by the Pennsylvania Historical and Museum Commission, Titusville. The 1864 report was by B. F. Isherwood and is cited as "Report of the Secretary of the Navy for 1864, House Executive Document, Number 1, 38th Congress, 2nd Session, page 1096, Government Printing Office, Washington, 1864." The second report, also by Isherwood, is cited as "Report of the Secretary of the Navy for 1867, House Executive Document, Number 1, 40th Congress, 2nd Session, pages 173–175. Government Printing Office, Washington, 1867." The much more positive reports about oil as a potential fuel for navy ships came from *New York Times* articles that were reprinted in the *Venago Spectator* on June 28, 1867, and the *Titusville Morning Herald* on July 10, 1867. The words of the secretary of the navy in 1920 come from "Fuel for the Navy," an extract of a speech made by Secretary of the Navy Josephus Daniels on December 18, 1920, published in *Journal of the American Society for Naval Engineers* 33, no. 1 (February 1921): 60–63.

The comments about the oil-fired fire engine came from an article in the *Boston Traveler*, which was reprinted in the *Venago Spectator* on October 11, 1867, and then reprinted again in the aforementioned *Pennsylvania Petroleum, 1750–1872*.

Most of Alaska's oil is shipped to Washington and California. A small amount is shipped by tanker to the refinery near Nikiski, Alaska. Alaskan oil is not exported to foreign refiners.

Vern Miller of Tesoro was the refinery engineer who talked to me about refineries. We were accompanied by Kip Knutson, also of Tesoro. After having had my requests for refinery tours turned down by several other energy companies, I found Tesoro's openness and hospitality refreshing. Tesoro hosts an excellent explanation of the refining process online at www.tsocorp.com/Refining101/index.html.

The KFC firewalking incident is well known. See www.theage.com.au/articles/2002/02/27/1014704967158.html. I do not know of any explanation for the prevalence of reported burns associated with this incident.

The brief history of the modern firewalking movement was reconstructed from a number of sources, including lecture notes posted at www.people.vcu.edu/~dbromley/undergraduate/spiritualCommunity/lecturematerialspirit.html and information from the Firewalking Institute of Research and Education. The 1977 *Scientific American* article was by Jearl Walker in his "Amateur Scientist" column, published in August on page 126. Emily Edwards's "Firewalking: A Contemporary Ritual and Transformation" appeared in *Drama Review* 42, no. 2 (1998): 98–114. According to the journal's publishers, the *Drama Review* "focuses on performances in their social, economic, and political contexts."

CHAPTER 6: STEAMING MOUNTAINS

The Henry Stephens Washington quotation comes from "Petrology of the Hawaiian Islands: II, Hualalai and Mauna Loa," *American Journal of Science*, series 5, 6 (1923): 100–126. Among other accomplishments, Washington wrote a book, *Chemical Analyses of Igneous Rocks* (1903), containing more than eight thousand analyses of rocks. With three other scientists, he developed a detailed classification system for igneous rocks known as the CIPW Norm and still referred to today by geologists.

Mark Twain's mention of "Pele's furnaces" came from an article in the *Sacramento Daily Union*, November 16, 1866. In the same article, Twain complimented the Volcano House hotel, saying that "it seems to me that to leave out the fact that there is a neat, roomy, well furnished and well kept hotel at the volcano would be to remain silent upon a point of the very highest importance to any who may desire to visit the place." The article was one of a series of "letters" Twain wrote as a correspondent, essentially on sabbatical from his work as a journalist in the southwestern United States, before he published his first book. Even as a journalist, he was unmistakably Mark Twain, who, in his youth, retained a sense of adventure but also knew how to relax. "It has been six weeks since I touched a pen," he wrote during his journey. "In explanation and excuse I offer the fact that I spent that time (with the exception of one week) on the island of Maui. . . . I went to Maui to stay a week and remained five. I had a jolly time. I would not have fooled away any of it writing letters under any consideration whatever." The articles from Hawaii were republished as *Mark Twain's Letters from Hawaii*, ed. A. Grove Day (Honolulu: University of Hawaii Press, 1975), and remain readily available today.

Entries in the Volcano House registry have been collected and published in the cleverly illustrated collection *On the Rim of Kilauea: Excerpts from the Volcano House Register, 1865–1955*, ed. Darcy Bevens (Hawaii National Park: Hawaii Natural History Association, 1992).

The quotation about Pele on a house-hunting expedition comes from *Time* magazine, April 22, 1940.

Captain James Cook's journals have been reprinted under the title *James Cook: The Journals* (New York: Penguin, 2003). The Penguin Classic version includes commentary by Philip Edwards. Regarding the spread of diseases to Hawaii, Cook wrote, "As there were some venereal complaints on board both the Ships, in order to prevent its being communicated to these people, I gave orders that no Women, on any account whatever were to be admitted on board the Ships, I also forbid all manner of connection with them, and ordered that none who had the venereal upon them should go out of the ships. But whether these regulations had the desired effect or not time can only discover." Cook clearly realized that he could not keep men and women apart. He also understood that in-fection could be transmitted even when its outward signs were not visible.

My contact at the Hawaiian Volcano Observatory was Dr. Matt Patrick. He was kind enough to take time away from his work to answer my questions about volcanoes in general and Kilauea in particular. He has worked on volcanoes in Alaska, South America, Europe, and Hawaii. Appropriately enough for a man who loves volcanoes, he lives in the town of Volcano, close to the Hawaiian Volcano Observatory.

The Hawaiian Volcano Observatory publishes a weekly feature in the island's newspapers. The feature of August 28, 2008, "Foot-prints in Ka'u were probably made in 1790—but not by Keoua's party," says, "We measured 405 footprints to determine how tall the walkers were." The article concludes that most of the footprints were made by women and children, not warriors.

The term "shield volcano" was derived from the name of an Ice-landic volcano, called Skjaldbreiður, which means something like "broad shield." Although Mauna Loa is the largest known shield volcano on earth, Olympus Mons, on Mars, is even bigger, with a height of more than 15 miles and a width of 370 miles.

Of the fifty-seven people killed by the Mount St. Helens eruption, the remains of twenty were never recovered. HistoryLink.org, "The Free Online Encyclopedia of Washington State History," lists the names of the dead.

Dick Thompson's *Volcano Cowboys: The Rocky Evolution of a Dangerous Science* (New York: Thomas Dunne, 2000) offers a dra-matic and readable account of what it is to be a volcanologist. The book focuses to some degree on the Mount St. Helens eruption. Information about deaths during that explosion and Stanley Lee's words come from Thompson's book.

Charles Darwin's comments on volcanoes and the destruction at Concepcion come from the many letters he wrote to his second cousin William Darwin Fox. Anthony W. D. Larkum collected many of Darwin's letters in *A Natural Calling: Life, Letters, and Diaries of Charles Darwin and William Darwin Fox* (Springer, 2009). Darwin also seemed to recognize at least aspects of the reality of plate tectonics in "On the Connexion of Certain Volcanic Phenomena in South America; and on the Formation of Mountain Chains and Volcanoes, as the Effect of the Same Powers by Which Continents Are Elevated," *Transactions of the Geological Society of London* 5, no. 3 (1840): 601–31. He wrote: "The whole western part of the continent [South America] has been almost simultaneously affected, it appears to me, that there is little hazard in assuming, that this large portion of the earth's crust floats in a like manner on a sea of molten rock. Moreover,—when we think of the increasing temperature of the strata, as we penetrate downwards in all parts of the world, and of the certainty that every portion of the surface rests on rocks which have once been liquefied;— when we consider the multitude of points from which fluid rock is annually emitted, and the still greater number of points from which it has been emitted during the few last geological periods inclusive, which, as far as regards the cooling of the rock in the lowest abysses, may probably be considered as one, from the extreme slowness with which heat can escape from such depths;— when we reflect how many and wide areas in all parts of the world are certainly known, some to have been rising and others sinking during the recent era, even to the present day, and do not forget the intimate connexion which has been shown to exist between these movements and the propulsion of liquefied rock to the surface in the volcano;—we are urged to include the entire globe in the foregoing hypothesis."

Melville's description of Nukuheva comes from *Typee*, published in 1846. *Typee* was Melville's first and most successful (during his lifetime) book. The book was published as nonfiction, although today it is widely believed to have been a fictionalized account based on Melville's experiences in the Marquesas Islands. Like many whalers, Melville jumped ship before finishing his entire multiyear cruise. The whale ship was the *Acushnet* from New Bedford. His point of disembarkation was Nukuheva, today commonly spelled Nuku Hiva, at 127 square miles the largest of the Marquesas Islands. He remained in Nukuheva for three weeks before sailing on another whale ship bound for Hawaii.

The bombings of Mauna Loa lava flows in 1935 and 1942 are described by many authors. One description can be found in J. P. Lockwood and F. A. Togerson, "Diversion of Lava Flows by Aerial Bombing—Lessons from Mauna Loa Volcano, Hawaii," *Bulletin of Volcanology* 43, no. 4 (1980): 727–41. Other attempts at lava diversion in Hawaii have met with equally poor results. In contrast, John McPhee's well-known book *The Control of Nature* (New York: Farrar, Straus and Giroux, 1989) describes an apparently successful (in the short term, at least) attempt to divert an Icelandic lava flow that threatened the destruction of Vestmannaeyjar harbor, an important resource (important, that is, to those who lived on the Icelandic island of Heimaey). The diversion was accomplished by pumping salt water onto the flows, selectively cooling the lava to create a levee that sent the lava elsewhere. Another description can be found in the 1983 United States Geological Survey report *Man against Volcano: The Eruption on Heimaey, Vestmannaeyjar, Iceland*, by Richard Williams and James Moore, available online at http://pubs.usgs.gov/gip/heimaey/heimaey.pdf.

Charles Keeling described his work in "Rewards and Penalties of Monitoring the Earth," *Annual Review of Energy and the Environment* 22 (1998): 25–82. Over time, Keeling's work became well known and arguably forms the foundation—or at least part of the foundation—for current thinking about climate change, but his article clearly describes the challenges of maintaining long-term funding for monitoring the atmosphere. Throughout his career, his funding stream was threatened and frequently cut off, often because it was considered to be "monitoring" rather than "research." In one episode, a National Science Foundation representative wrote, "I believe that we in this program must tread a narrow line between that work which constitutes basic research and that work which constitutes fairly routine monitoring." Keeling's work, in fact, could hardly have been considered routine. There were real and ongoing challenges with calibration of measurements and with odd oscillations in the long-term patterns that he detected. And, of course, one can argue that understanding the atmosphere and the earth's climate would require decades of basic research, even if that basic research was nothing more than the repeated collection of measurements over many years. The difficulty of maintaining funding for long-term monitoring remains a tremendous challenge in the environmental sciences.

The man who showed me around the Mauna Loa Observatory was John Barnes, the facility director. When I arrived, he had just finished talking to a class of Cornell students. He talked to me, my companion, and a retired couple, spending more than an hour showing us around. The laboratory, though somewhat remote, has increasingly attracted global-warming tourists. John's enthusiasm and passion for his facility was nothing short of inspirational.

Svante Arrhenius's comments about his climate change calculations come from "On the Influence of Carbonic Acid in the Air upon the Temperature of the Ground," *London, Edinburgh, and Dublin Philosophical Magazine and Journal of Science* 41, no. 251 (1896): 238–76. His paper is often cited as a harbinger of today's concerns about climate change, but in fact he was very much focused on understanding past changes in climate.

Guy Callendar wrote a number of papers, but I have quoted here from "The Artificial Production of Carbon Dioxide and Its Influence on Temperature," *Quarterly Journal of the Royal Meteorological Society* 64 (1938): 223–40. In addition to the benefits he associated with warming, he also mentioned that increased carbon dioxide levels would improve plant growth. This is true for some plants but not for others—that is, some plants cannot process the increased concentrations of carbon dioxide currently available, and others can process the concentrations of carbon dioxide currently available and could process even higher concentrations. It seems likely that plants capable of processing more carbon dioxide will have a competitive advantage over those that do not, and this advantage may eventually lead to population explosions of certain species. In fact, this may already be occurring. For example, from an article by Jack A. Morgan and his coauthors in 2007: "A hypothesis has been advanced that the incursion of woody plants into world grasslands over the past two centuries has been driven in part by increasing carbon dioxide concentration in Earth's atmosphere. Unlike the warm season forage grasses they are displacing, woody plants have a photosynthetic metabolism and carbon allocation patterns that are responsive to CO_2, and many have tap roots that are more effective than grasses for reaching deep soil water stores that can be enhanced under elevated CO_2." The full citation is J. A. Morgan, D. G. Milchunas, D. R. LeCain, M. West, and A. R. Mosier, "Carbon Dioxide Enrichment Alters Plant Community Structure and Accelerates Shrub Growth in Shortgrass Steppe," *Proceedings of the National Academy of Sciences* 104, no. 37 (2007): 14724–29.

In the very brief summary of the history of climate change research, I skipped many important names and important papers for the sake of brevity and (I hope) to avoid sending readers to sleep, the liquor cabinet, or the television. Among these are E. O. Hurlbert's work in 1931, Lewis D. Kaplan's work in 1952, Nelson Dingle's work in 1954, Hans Seuss's work in 1955, Gilbert N. Plass's work in 1956, Willard Libby's work in the 1950s, work by various weapons engineers interested in atmospheric physics during and after World War II, and others. I did not even mention the importance of the development of digital computers, a critical tool needed to understand the complexities of climate change. For readers interested in more information, Spencer Weart offers a remarkable account in *The Discovery of Global Warming* (Cambridge: Harvard University Press, 2003), which he has subsequently updated. He also maintains an exhaustive (but not exhausting, thanks to his masterful use of links, which allow readers to pursue their own interests at their own pace, rather than being held captive by the author's interests and pace) website on the topic at www.aip.org/history/climate/index.htm. Weart was trained as a physicist but is known for his work as a historian of physics and geophysics. He worked as director of the Center for History of Physics of the American Institute of Physics in College Park, Maryland, before his retirement in 2009. Anyone who aspires to write a book about the history of climate change science, take heed: Weart's effort will be hard to beat.

Maunder Minimums and the possibility that they impact climate were famously described in John Eddy's paper "The Maunder Minimum," *Science* 192, no. 4245 (June 1976): 1189–1202.

The description of the cooling of the lava lake at Kilauea Iki comes from an article published by the Hawaiian Volcano Observatory on January 9, 2003, "Kilauea Iki's Lava Lake Has Finally Crystallized." The article is available online at http://hvo.wr.usgs.gov/volcanowatch/2003/03_01_09.html.

Garrett Smathers and Dieter Mueller-Dombois studied plant colonization of the Kilauea Iki crater. They described their findings in a highly accessible manner in *Hawai'i, the Fires of Life: Rebirth in Volcano Land* (Honolulu: Mutual Publishing, 2007). The two men, along with a host of assistants and students, watched the crater recolonize for five decades before publishing this book.

The chirpless cricket of Hawaiian lava flows, also called a lava cricket or lava flow cricket, was first described in the scientific literature in 1978, when it was named *Caconemobius fori*. Another cricket within this genus lives in lava caves. Francis Howarth described trapping lava crickets in "Neogeoaeolian Habitats on New Lava Flows on Hawaii Island: An Ecosystem Supported by Windborne Debris," *Pacific Insects* 20, nos. 2–3 (1979): 133–44.

CHAPTER 7: BOOM

Reliable and consistent measurements or estimates of the core temperature of hydrogen bombs do not seem to be available. Temperatures vary quickly over time and space. The estimated twenty million degrees used here was based on extrapolation between a number of estimates from different sources.

President Eisenhower's Atoms for Peace speech can be found in whole at www.atomicarchive.com/Docs/Deterrence/Atomsfor-peace.shtml.

The Red Dog Mine is one of the world's largest zinc producers. Red Dog has been in operation since 1989, when it was developed through an agreement between an Inupiat Alaskan corporation and a multinational mining corporation. Partly processed ore is stored on-site in winter and shipped out in summer, when the ice clears from the Chukchi Sea. Despite Red Dog's stated commitment to environmental stewardship, the operation has attracted criticism in part over concerns about soil contamination.

Destructive power in nuclear weapons is often expressed in kilo-tons or megatons—that is, destructive power is expressed as the weight of an amount of TNT that would be required to generate a similar effect. Kilotons are thousands of tons, and megatons are millions of tons of TNT. A one-hundred-kiloton shot, such as the Sedan shot, has the explosive power of one hundred thousand tons of TNT.

It is all but impossible to read about the development of nuclear weapons without coming across quotations and stories by and about Edward Teller. No matter what one thinks of nuclear weapons, Teller is an undeniably interesting character and a man whose intelligence and drive (as well as his often dark obsessions) are evident even in the shortest of statements by him and about him. I encountered many of the quotations and stories used in *Heat* in Gerard J. DeGroot's wonderful book *The Bomb: A Life* (Cambridge: Harvard University Press, 2004). Another very important source was Dan O'Neill's brilliant book *The Firecracker Boys* (New York: St. Martin's Griffin, 1994), which also introduced me to, educated me about, and eventually drew me to Point Hope and

Cape Thompson. *The Firecracker Boys* was especially important as a secondary source for Teller quotations about Project Chariot and Alaska. O'Neill and many other journalists suggest that Teller's interest in Plowshare projects was driven by his desire to test bombs, but Teller, like most people, was driven by a complex set of motivations. O'Neill also suggests that Teller knew that there were no funds for harbor development other than those that would support the blasting work, but it is possible that Teller assumed appropriate public works funding would be forthcoming once the geography of the harbor was in place.

Information on the bombing of Hiroshima is, of course, widely available. John Hershey's short but powerful *Hiroshima* (1946; republished by Vintage Books in 1989), based on interviews of survivors, offers a particularly horrific viewpoint. I found it difficult to research and write about the bombing of Hiroshima. It is impossible to separate an intellectual understanding of World War II and a technical understanding of the atomic bomb from the horror that occurred when the bomb was dropped on a city.

Edward Teller's *Memoirs: A Twentieth-Century Journey in Science and Politics* (New York: Basic Books, 2001), written by Teller with Judith L. Shoolery, provides fascinating insights to a brilliant but complicated man who lived through very interesting times. Teller's memoirs do not express a tremendous interest in, or concern for, environmental issues. Also, for a man of his brilliance and interests, his knowledge of basic northern biology is appalling. A footnote in his memoir reads, "Work on the North Slope [oil fields] was flourishing and so were the caribou, which had multiplied impressively near human habitation. Because the caribou's two enemies—fires and polar bears—were limited near the settlements, the caribou had benefitted." Although it is true that caribou numbers increased in the years following development of the Prudhoe Bay oil fields, Teller's belief that fire was an important enemy of the caribou was incorrect, as was his belief that polar bears—which in general live on the sea ice, feeding on seals and seldom encountering caribou—are an important enemy of caribou.

G. Peter Kershaw, Peter A. Scott, and Harold E. Welch assessed the heat value of seal oil lamps (*kudliks*) in traditional snow igloos in their article "The Shelter Characteristics of Traditional-Styled Inuit Snow Houses," *Arctic* 49, no. 4 (1996): 328–38. According to the authors, a seal oil lamp fueled by the fat of a single seal (presumably a ringed seal, *Phoca hispida*) could heat a small igloo for 6.3 days, while a larger igloo required the fat of a single seal every 3.7 days. Outside temperatures during the experiment were around minus thirty-five degrees Fahrenheit.

Eric Jay Dolin's *Leviathan: The History of Whaling in America* (New York: W. W. Norton, 2007) offers a wonderful history of whaling. Among other things, Dolin describes the influence of whaling on the opening of Japan to the West. In 1845, three years before Captain Roys sailed north of the Bering Strait, a Sag Harbor whaler called the *Manhattan,* under the command of Mercator Cooper, was among the first whale ships to sail to Japan. Although Japan was a forbidden coast at the time, Cooper had stumbled on eleven shipwrecked Japanese sailors at St. Peter's Island and felt justified in returning them to their homeland. The Japanese graciously welcomed Cooper, but they also welcomed his departure. Eventually interactions with whalers—some of them shipwrecked on Japanese shores and then imprisoned—contributed to Commodore Matthew Perry's 1853–54 expedition to Japan and the subsequent treaty that opened Japan to the West. In 1851, Herman Melville wrote, "If that double-bolted land, Japan, is ever to become hospitable, it is the whale ship alone to whom the credit will be due." Melville, of course, does not comment on the Gokoku Shrine—the shrine that was built in the memory of those who died in a civil war related to the opening of Japan, and the shrine that became ground zero for the atom bomb that was dropped on Hiroshima.

Oil from the sperm whale was a superior product for lighting—it burned a clean white light, almost smokeless, bright enough for use in lighthouses and sometimes said to have lit the lamps of the Enlightenment. In contrast, oil from bowhead whales, sometimes sold as "brown oil," did not burn as well, but by the 1860s bow-heads were easier to find than the heavily hunted sperm whales. In addition, the bowhead, while a large and potentially dangerous whale, would have been easier and safer to hunt than the sperm whale.

The quotation "They don't like the cold iron," as well as the anecdote about whales responding to the sounds of whaleboats against ice, was reported in J. R. Bockstoce, *Whales, Ice, and Men: The History of Whaling in the Western Arctic* (Seattle: University of Washington Press, 1986).

The captain's comment about the rapid decimation of bowhead stocks came from an anonymous letter written in 1853 and reported in the *Whaleman's Shipping List* (New Bedford, Mass.), May 3, 1853. It was mentioned in the report by J. R. Bockstoce, D. B. Botkin, A. Philp, B. W. Collins, and J. C. George, "The Geographic Distribution of Bowhead Whales, *Balaena mysticetus,* in the Bering, Chukchi, and Beaufort Seas: Evidence from Whaleship Records, 1849–1914," *Marine Fisheries Review,* 2005. The comment on bowhead stocks comes from an 1852 journal entry mentioned in the report. The journal is in the Old Dartmouth Historical Society Collection (no. 145) at the New Bedford Whaling Museum, New Bedford, Mass.

Jörg Freidrich's *The Fire: The Bombing of Germany, 1940–1945* (New York: Columbia University Press, 2002) describes the impact of firebombing on Germany during World War II. The author also narrates part of the 2003 documentary *Firestorm: The Allied Bombing of Nazi Germany* (directed by Michael Kloft), which offers a sickening glimpse of the realities of firebombing.

Early versions of jellied gasoline used rubber as a thickening agent. Later, after the Japanese cut off supplies of rubber, a mixture of aluminum naphthenate and extracts from coconuts was used, leading to the name "napalm." Still later, new methods were developed using polystyrene mixed with gasoline and benzene, but "napalm" stuck. Napalm was first used in World War II in the bombings of Tinian, in the South Pacific, and later at Iwo Jima and Okinawa. Napalm was also used in the Korean War and the war in Vietnam.

Haney's words about the firebombing of Kobe come from his book *Caged Dragons: An American POW in WWII Japan* (Ann Arbor: Sabre Press, 1991). Ray "Hap" Halloran's memory of praying during the firebombing of Tokyo come from Mari Yamaguchi's article "Museum Recalls Tokyo Firebombing," Associated Press, March 10, 2002.

Obata Masatake was interviewed by the Australian Broadcasting Company for a 1995 *Radio Eye* episode called "Tokyo's Burning." The show was based in part on Robert Guillain's *I Saw Tokyo Burning* (New York: Doubleday, 1981). Guillain was a French journalist who lived in Japan from 1938 until after the end of the war.

Einstein's letter to President Roosevelt is readily available in its entirety. One source is http://hypertextbook.com/eworld/einstein .shtml#first.

The Nobel Prize is named after Alfred Nobel, a Swedish engineer and industrialist who invented and manufactured dynamite and other explosives. Dynamite is essentially a stable form of nitroglycerin. Although Nobel's dynamite was widely used in warfare, Nobel saw it primarily as a tool for civil engineering works, a view that superficially parallels Teller's purported position with regard to the Plowshare program. Although Nobel sometimes claimed to be a pacifist, and though it has been said that he was deeply troubled by a newspaper headline that described him as a "merchant of death," he knew that his products were being sold to warring nations and sometimes to both sides of conflicts. When Nobel died, his factories were producing 66,000 tons of explosives each year. Nobel's will stated that he wished the Peace Prize to go "to the person who shall have done the most or the best work for fraternity among nations, for the abolition or reduction of standing armies and for the holding and promotion of peace congresses." One can only imagine what it would be like to have tea with Teller and Nobel at the same table, discussing the peaceful use of explosives.

Teller's statement about looking "straight at the bomb" appears in several sources, including John Langone, "Edward Teller: Of Bombs and Brickbats," *Discover*, July 1984, 62–68.

The comment from Herbert York about his recollection of Teller explaining how a hydrogen bomb might work was reported in Gerard J. DeGroot's aforementioned *The Bomb: A Life* (Cambridge: Harvard University Press, 2004), 175.

Although Mike, the first hydrogen bomb, would create temperatures far hotter than anything previously created by humans, its size was driven largely by the need to keep the deuterium—the heavy hydrogen that was the fuel for fusion in the Mike device—at very low temperatures.

The description of Greek fire as coming on "as broad in front as a vinegar cask" came from the thirteenth-century memoir of Jean de Joinville, a French knight and crusader. Ethel Wedgwood's *The Memoirs of the Lord of Joinville: A New English Version* (New York: E. P. Dutton, 1906) is widely available, including on Google Books, and offers a firsthand description of life during the Crusades.

Edward Teller's wonderful remark implying that he could move mountains—worded in a manner that made it suitable for greeting cards or advertising copy—comes from the *Anchorage Daily Times*, June 26, 1959. I first came across this quotation in Dan O'Neill's *The Firecracker Boys* (New York: St. Martin's Griffin, 1994).

Ship's surgeon John Simpson served under Rochfort Maguire. His records include heights and weights of a number of Point Barrow residents—from five feet one inch to five feet nine and one-half inches, and from 125 to 195 pounds. His writing is included in Maguire's journal as an appendix.

The description of the colors in flames was based on a description and illustrations in John W. Lyons, *Fire* (New York: Scientific American Books, 1985). The book is well laid out and includes brilliant illustrations.

Nathan Hodge and Sharon Weinberger's *A Nuclear Family Vacation: Travels in the World of Atomic Weaponry* (New York: Bloomsbury, 2008) reassured me that I am not the only author in the world writing rather odd books. It is well worth reading.

Einstein's famous equation relates energy (E), mass (M), and a constant representing the speed of light (C). The equation appeared in Einstein's 1905 paper "Does the Inertia of a Body Depend upon Its Energy-Content?" $E = MC^2$ tells us that matter and energy are different expressions of the same thing. In practice—that is, to apply the formula with compatible units—energy can be expressed in ergs as $g \cdot cm^2/sec^2$, mass can be expressed in grams, and the speed of light can be expressed in centimeters per second. One erg is the same as 1 dyne per centimeter or 100 nanojoules. The speed of light squared becomes 900 quintillion cm^2/sec^2. This means that the formula could be re-expressed as $E = M \times 900$ quintillion. Clearly, relatively small masses hold tremendous amounts of energy. In a recorded lecture (see www.youtube.com/watch?v=CC7Sg41Bp-U), Einstein suggested that the repercussions of his formula—the realization that matter and energy are different manifestations of the same thing—presented "a somewhat unfamiliar conception to the average mind."

It is often reported that the jeep was originally called the General Purpose Vehicle, which inevitably became the GP, which inevitably became the jeep. Others claim that the iconic name came from a comic strip, and occasionally it is attributed to a test driver's quip that was quoted in a syndicated column. Ernie Pyle's words about the Coleman stove and the jeep come from a Coleman website, www.coleman.com/coleman/colemancom/newsrelease.asp?releasenum=338.

Teller's statement about the Soviets setting off nuclear weapons in the name of civil works was reported in Dan O'Neill's *The*

Firecracker Boys (New York: St. Martin's Griffin, 1994). O'Neill attributes the statement to a tape recording of Teller's Strategic Defense Initiative speech to Commonwealth North at the Captain Cook Hotel, Anchorage, Alaska, June 9, 1987, recorded by Chris Toal of SANE Alaska.

The exact number of bombs used for nonmilitary purposes in the former Soviet Union is not widely known. Milo D. Nordyke's report "The Soviet Program for Peaceful Uses of Nuclear Explosions," *Science and Global Security* 7 (1998): 1–117, suggests that 128 nuclear explosives were used, while Gerard J. DeGroot's *The Bomb* (Cambridge: Harvard University Press, 2004) puts the number at 156. Most of the information on the Soviet Program for Peaceful Uses of Nuclear Explosions comes from Nordyke's article, as well as a report by Nordyke with the same title, published by the Lawrence Livermore National Laboratory on July 24, 1996 (UCRL-ID-124410). The article and report were based on a number of articles and books published by Russians after Glasnost.

"Chetek," in the International Chetek Corporation of Moscow, is an acronym from Russian words for man, technology, and capital.

The chairman of the meeting at which the Chetek Corporation announced its plans to market nuclear explosions for waste management was John M. Lamb, and his comments were reported by William J. Broad in "A Soviet Company Offers Nuclear Blasts for Sale to Anyone with the Cash," *New York Times*, November 7, 1991. Somewhat ironically, the meeting was sponsored by the Canadian Center for Arms Control and Disarmament. Despite Lamb's comment, at least some Western scientists took the Chetek Corporation's suggestions seriously. A Lawrence Livermore National Laboratory scientist, also quoted in the *New York Times* article, said that the Chetek approach would be the cheapest way to dispose of hazardous materials, including unwanted nuclear warheads.

Edward Teller's memoir offers few insights on his visits to Cape Thompson and what he did there. I could not determine if

Teller used the radio whose skeleton still remains at the site, or if he rode the Weasels that were abandoned there, or if he ate from trays like the one that I found on the ground near my tent. I can only guess that some of what remains at Cape Thompson is from later cleanup crews, but certainly some of the equipment must have been present when Teller was working on the project.

The statements by the Point Hope Village Council, Kitty Kinneeveauk, and Joseph Frankson come from Dan O'Neill's *The Firecracker Boys* (New York: St. Martin's Griffin, 1994). O'Neill used a number of sources for these statements, including tape recordings of public meetings.

CHAPTER 8: THE TOP OF THE THERMOMETER

Herschel's words about the sun as "the ultimate source of almost every motion" come from his *A Treatise on Astronomy* (1833), republished in 2005 as an Elibron Classic, by Adamant Media Corporation. The passage on the sun as "the ultimate source" has been repeated many times. For example, it appeared in vol. 13 of *All the Year Round: A Weekly Journal, Conducted by Charles Dickens*, in 1865. Herschel's speculations about life on the sun and on other planets are less often repeated. The passages quoted here come from his paper "On the Nature and Construction of the Sun and Fixed Stars," *Philosophical Transactions of the Royal Society of London* 85 (1795): 46–72.

The physicist Hans Bethe is usually recognized as the first to realize that the energy of stars and the sun came from fusion. His paper "Energy Production in Stars," *Physical Review* 55 (1939): 434–56, was published just thirteen years before the first fusion bomb—that is, the first hydrogen bomb, Teller's Super—was exploded on Elugelab Island, Eniwetok, and only twenty-three years before the Sedan shot in the Nevada desert.

When discussing the evolution of stars—when thinking about what happened billions of years ago and what will happen billions of years from now—it is natural to question authority. How do we know this stuff? From Arthur Stanley Eddington, in his book *Stars and Atoms* (Oxford: Clarendon Press, 1927), answering this question in the context of his work related to the star Sirius and its orbiting companion, a white dwarf: "We learn about the stars by receiving and interpreting the messages which their light brings to us. The message of the Companion of Sirius when it was decoded ran: 'I am composed of material 3,000 times denser than anything you have ever come across; a ton of my material would be a little nugget that you could put in a matchbox.' What reply can one make to such a message? The reply which most of us made in 1914 was—'Shut up. Don't talk nonsense.'"

Loring Danforth's *Firewalking and Religious Healing* (Princeton University Press, 1989) is an academic work describing firewalking from the perspective of the author, who spent time in Greece studying traditional firewalkers and eventually walked through fire himself. The words of the man who said he believed he could survive a nuclear explosion were attributed to a *Rolling Stone* article by Jon Krakauer published in 1984.

Teller's memoir mentions Fermi more than eighty times. The two men, from a young age, were friends, and both had fled Europe to the United States to avoid fascism. But after the war, Fermi objected to the development of thermonuclear weapons. "A desirable peace," he wrote, "cannot come from such an inhuman application of force." In 1954, before his fifty-fourth birthday, Fermi died from cancer caused by exposure to radiation. In 1962, Teller received the prestigious Enrico Fermi Award, in part for his "efforts to strengthen national security."

The physicist I interviewed about supercolliders and who later showed me around at the Brookhaven laboratory was Barbara Jacek. Her knowledge of particle physics was astounding, especially when mixed with her willingness to spend time with a writer and her understanding of the need to make extremely complicated technical information accessible to nonspecialists.

Protons and neutrons are two kinds of hadrons known to most people. The lesser-known hadrons include antiprotons, omegas, lambdas, and kaons.

The Kelvin scale uses degree increments equivalent to those of the Celsius scale, but in Kelvin zero is set at absolute zero. At four trillion degrees Kelvin, one can ignore the difference between Kelvin and Celsius—at four trillion degrees, the small difference in the zero points becomes irrelevant.

My firewalking instructor was Claudia Weber (http://hypnotherapy-chico.com/FIREWALKwithClaudia.en.html). Her enthusiasm for firewalking and for the human spirit in general is inspiring. Firewalking has been called a New Age fad, a sham, and worse. It has been the subject of a tremendous amount of information and misinformation. A certain contingent seems to think that the scientific explanation of firewalking somehow decreases its value, and perhaps for some people it does. But for the vast majority of people, including the dozens of firewalkers whom I spoke to in the course of writing this book, firewalking offers a unique and positive experience of a very personal nature. It is interesting to watch or listen to scientists who set out to debunk firewalking with scientific explanations. They generally walk through fire themselves to show that it can safely be done without a mystical state of mind, and in so doing they seem to get a tremendous thrill from the act itself.

Dixy Lee Ray, a 1970s environmentalist who endorsed the benefits of nuclear energy, was not afraid of controversy. She believed in science and in the power of science to inform public policy. Her book with Lou Guzzo, *Environmental Overkill: Whatever Happened to Common Sense?* (New York: HarperCollins, 2000), criticized the environmental movement for, in her view, ignoring scientific evidence that did not support predetermined political positions. "We are aware," she wrote, "that some may conclude that we are 'anti-environment.' That would be wrong." She also believed that it would be wrong "to downplay the remarkable resilience and re-covery powers of nature." In the first chapter, she questions the science behind climate change and suggests or at least implies that the Rio Earth Summit and its statements regarding the need to reduce greenhouse gas emissions at the expense of developed nations had more to do with redistribution of wealth than with environmental protection. She claims that the vice chairman of the Rio Earth Summit openly admitted that the conference agenda "was based upon the International Socialist Party's platform." *Environmental Overkill* is now more than ten years old, and it is impossible to say if the late Professor Ray would have changed her views about climate change as more data have become available.

The collider at Brookhaven is 2.4 miles in circumference and can, according to some sources, generate gold ion collisions with a power of 40 tera electron volts. The collider near Geneva, Switzerland, the Large Hadron Collider (LHC), has a circumference of 17 miles and can, according to some sources, generate lead ion collisions with a power of 574 tera electron volts.

Like Tyndall in his biography of Faraday, Enrico Fermi, in an address called "The Future of Nuclear Physics" (Rochester, N.Y., January 10, 1953), also commented on the value of basic research, such as the research that would be done a half century later at Brookhaven: "History of science and technology has consistently taught us that scientific advances in basic understanding have sooner or later led to technical and industrial applications that have revolutionized our way of life. It seems to me improbable that this effort to get at the structure of matter should be an exception to this rule."

INDEX

ABOUT THE AUTHOR

Bill Streever is the author of the national bestseller *Cold*. He lives in Alaska.